ABIDING IN GOD

A Scientist's Insights
into Our Guiding Spirits,

Inspired by the Courtship Letters
of His Parents

Mark Johnson

Abiding in God © 2024 Mark Johnson

All rights reserved. No part of this publication may be reproduced or transmitted in any form or by any electronic or mechanical means including photo copying, recording, or any information storage and retrieval system now known or to be invented, without permission in writing from the publisher or the author.

Name: Johnson, Mark
Title: *Abiding In God* by Mark Johnson
Interior and cover layouts: Robert Ousnamer

ISBN: 978-1-963611-93-9
LCCN: Applied for

Subjects: 1. Books > Religion & Spirituality > Religious Studies > Science & Religion
2. Books > Christian Books & Bibles > Christian Denominations & Sects > Protestantism > Inspirational
3. Books > Parenting & Relationships > Marriage & Adult Relationships

Published by EABooks Publishing
a division of Living Parables of Central Florida, Inc.
eabookspublishing.com

Opinions expressed in this book are those of the author and may or may not be representative of EABooks Publishing.

First Impressions of *Abiding in God*

What a charming collection of family love letters! . . . The strong Christian faith expressed in the letters is a familiar reflection of the European piety movement's major influence in Sweden prior to the great wave of immigration to the Midwestern U.S. in the late 1860s. I lived that too.

Janel Palmquist, student and beneficiary of our Swedish-American Christian legacy

This memorable book is a true love story between two young people separated by miles in the rural Midwest during the 1930s. . . . The read is easy and meaningful, as Christian family life is explored.

Pamela Chally, Professor and Dean Emeritus, University of Florida

What a gift . . . Parents always retain some degree of mystery to their children, even when we're close to them, but being able to read their letters, particularly from their young adulthood, reveals new aspects of what Mark Johnson so aptly calls their expressed spirits . . . Mark's references to George Mead's Mind, Self, and Society have inspired me to read it . . . One day, I'd love to hear more about his extraordinary experience of contemplating mind and having luminous fibers shoot from his midsection. . . . Some Quakers have spoken of similar experiences.

Scott Matthews, Professor of History, Florida State College at Jacksonville

I found myself in a state of disbelief and wonder. How could an ... ordinary farmer and an ... ordinary schoolteacher living separately on the Midwestern prairie in 1937 create a collection of letters to each other that are literally masterpieces of Christian love? ... Your heart and your soul will be touched so beautifully as you read these letters of love and spiritual communion .

David Brown, Retired School Psychologist.

Our contemporary society misses so much in their own relationships compared to Walter and Margaret. The letters ... express their mutual faith that God is intimately involved in ... their relationship and ... the world. In this context, Mark Johnson deeply examines what faith means for him.. ... His analysis ... encompasses philosophical and spiritual aspects which challenge readers to contemplate in their own faith journey. ... Read this book first for its story and then again for its meaning.

Constance Hastings,
The Trouble with Jesus: Considerations Before You Walk Away

CONTENTS

PREFACE -- ix

INTRODUCTION --- xi

1. HELP WITH LIFE ON THE FARM ------------------------------------- 1

    The May 7–May 17 Letters

    A Glimpse Back -- 34

    The Night of the Little Self -------------------------------------- 36

2. I DON'T MEAN TO BE INDEPENDENT OF YOU ------------------------- 42

    The May 17–May 25 Letters

    A Glimpse Back -- 68

    A Helper Fit for Me -- 69

3. DR. PIHLBLAD -- 75

    The May 27–June 15 Letters

    A Glimpse Back --- 101

    A Trip with Walter and Margaret -------------------------------- 103

4. I'M ON MY WAY -- 109

    The June 16–July 4 Letters

A Glimpse Back ---125

Watching the Expressed Spirits at Play in These Letters ----125

5. GOD'S HARD WORKING---131

The July 5–July 17 Letters

A Glimpse Back ---146

6. A LITTLE SWELLING ON MY FOOT -------------------------------------148

The July 19–July 29 Letters

A Glimpse Back ---165

7. YOU STILL LOVE ME THO, DON'T YOU? -------------------------------167

The August 15–September 10 Letters

A Glimpse Back ---213

8. JOY IN THIS LIFE AND THE LIFE HEREAFTER --------------------------217

What Lives On ---220
Our Most Basic Spiritual Choices--------------------------------224
Consequences of the Elemental Guiding Spirits We Heed-238
Assuring Life in the Hereafter ----------------------------------291

AFTERWORD ---311

A CLOSING REFLECTION AND INVITATION -------------------319

ACKNOWLEDGMENTS --322

Appx 1: A Tally of Walter and Margaret's Personal Concerns ----325

Appx 2: Expressions of Margaret's and Walter's Basic Spirits ----328

Appendix 3: Portrayals in the Gospels of Our Guiding Spirits ----336

Meet Mark--347

PREFACE

In his prelude to *Earth-Honoring Faith: Religious Ethics in a New Key*, Larry L. Rasmussen suggests that human beings must start singing a new song if humanity is to survive on Earth. I've yet to read the book, but seeing the ancient in the new is central to understanding the different voices in this final volume of a trilogy on the courtship letters of my parents, Walter and Margaret.

We cannot help but sing the songs of our lives. They become duets when we fall in love. Only close friends and relatives are privy to these duets and then, only occasionally. That was the case for my parents until their courtship letters surfaced after they had passed away.

The courtship of which my parents sang came with three rich and ancient verses: getting to know each other, weighing the pros and cons of a permanent relationship, and preparing to publicly commit at the altar. It's a song that has been sung many ways. Their harmony is a symphony of love colored by their religious heritage, a heritage I share.

As the curator of these letters, I am joining in their song by coloring bits of what they sing with complementary reflections. This was all I did in the first volume, *Encountering God: Reflections on the Courtship Letters of My Parents*.

Inspired by things of which they sang in the second volume, *Questioning God: Philosophical Reflections on Pivotal Concerns in My Parents' Letters*, I broke out into three solos. I titled my songs

"Freeing Unifying Religious Truths from Their Partisan Entanglements," "Visions of the Oncoming World Order," and "Seeing Jesus: Same World, New Eyes." What I sang had already been sung by others, as noted in the extensive references.

    The songs I've been inspired to sing in this volume are largely my own compositions, evoked by what my parents are singing in their letters. Walter's questioning a friend's revelation leads into the first, "The Night of the Little Self." It is an attempt to describe a startling inversion of my sense of self that opened my eyes to the spirits at play in the world and in these letters. This is followed by "A Helper Fit for Me" when they describe what they truly enjoy in each other. "A Trip with Walter and Margaret" describes a trip my parents took with me to the places and families mentioned in these letters. "Watching the Embodied and Expressed Spirits at Play in These Letters" briefly zooms in on what transpired spiritually in a particularly interesting exchange of letters. How our guiding spirits must be prioritized if we are to do more than just survive on the blue marble we call Earth is proffered in the grand finale, "Joy in This Life and the Life Hereafter."

    To better understand how these guiding spirits infuse our lives and to more fully appreciate that of which I sing, it helps to have a documented case of these spirits at play in our daily lives and in the typifying time frames of those guiding spirits. The courtship letters of Walter and Margaret are notable in that regard. They are well written and cover a critical and familiar time in the lives of so many: the encountering, questioning, and committing phases leading up to a meaningful marriage. In this last third of their letters, we will be looking for what caused Margaret and Walter to sing during the last four months before they publicly declared their commitment at the altar.

INTRODUCTION

My parents, Walter and Margaret, met at an autumn church outing in 1937 when she visited her sister Evodia and her brother-in-law, Reuben, the pastor of Walter's church in Greeley, Colorado. Shortly afterward, Margaret returned to Washburn, North Dakota, to direct choral groups and teach high school English. Aside from two brief visits, theirs was a lettered courtship. The letters in this volume are the latter third of their exchanges in preparation for their wedding and their life together on a farm.

Initially, there were a few rough edges. Both Walter's parents had immigrated to America from Sweden, his dad as an indentured servant and his mom as an adventurous young woman. They spoke only Swedish at home. That changed when Walter, their first child, flunked the first grade because he couldn't speak English. He would skip two grades by the time he finished high school, but he didn't go to college. He sent his first letter to "Miss Margaret Leaf, Clarissa, Minnesota," her parents' address. In it, he apologizes for his grammar. He ends his letter in the hope of Margaret helpfully corresponding with his sister, promising to "ask her to write first if I get your address."

Margaret's brothers and sisters all went to Swedish Lutheran colleges. Her father, two of her brothers, and two of her brothers-in-law were pastors. Any challenges related to their educational and social backgrounds were overcome as a mutual vision quickly evolved of a heavenly father bringing them together, caring for them, and responding to their thoughts and prayers. This promising phase is covered in the first book of letters, *Encountering God: Reflections on the Courtship Letters of My Religious Parents* (*EG*). Meanwhile, world affairs were deteriorating economically and militarily. The possibility of another "great war" was unfolding. Walter was refused a needed farming loan. Margaret wondered why the livestock market had to "fluctuate so much."

These and other worries further materialized in the second book of letters in this series, *Questioning God's Will: Philosophical Reflections on Pivotal Concerns in My Parents' Letters* (*QGW*). Both had thoughts of end-times when Hitler's armies entered Austria. Walter wondered about God's will, and Margaret wondered about Walter's. Their winter letters ended with Walter feeling that "another, and possibly worse, depression is underway" and with Margaret hesitating to inform Mr. Thorson, the principal, that she wouldn't be coming back "if we don't plan to get married until a year or two."

In this third volume of letters, *Abiding in God: A Scientist's Insights into Our Guiding Spirits Inspired by the Courtship Letters of His Parents*, their intentions may vacillate, but those of nature don't. Spring is underway. Fields are planted, and the ground greens. Although Margaret may not see the writing on the wall, her colleagues do and enjoy teasing her about what life will be like on the farm. That teasing will pale against harbingers of what lies ahead for

Margaret when she learns she will be living next door to an alcoholic landlord and struggles with the details of a rapidly approaching wedding.

As in the first two books, the letters open stylistically with spiritual quotes, often augmented with personal contemplations. Postscripts inscribed along the edges of a letter are placed after the closing salutation. The numerous ellipses represent editorial deletions of mundane material. Misleading punctuation and confusing spelling errors have been corrected, but errors in contracting and capitalizing words that do not impede the reader are not. Some spelling errors, such as *hurriedly*, *potatoe*, and *lightening*, have been left in for authenticity. The goal is to convey how their thoughts and feelings were expressed at the time they were written and as they were written. It seems to have been an "in thing" for them to sometimes write *thot* for *thought*, *nite* for *night*, *altho* for *although*, and *brot* for *brought*.

Various editorial comments are interspersed throughout the letters. The shorter ones either explain an unfamiliar expression or briefly describe how the associated understanding played out in my parents' lives. They directly follow the relevant paragraph. Expressions underlined in the letters signal thought-provoking reflections to follow. These can be ignored by those caught up in the flow of the letters. Most signal shorter reflections at the end of the letter or in the Glimpse Back sections. A few are lead-ins to the titled reflections "sung" at the end of the chapter. Although the letters are historically relevant to these extended reflections, all can be meaningfully understood without having read the letters.

Biblical quotes are prevalent in both the letters and the editorial comments. No attempt has been made to ascertain the Bible versions used in the letters. Unless specified otherwise, my quotes are taken from the Revised Standard Version of my youth.

CHAPTER 1
HELP WITH LIFE ON THE FARM

The following are some of the 1938 headlines in the *New York Times*. May 4: "300,000 applaud Mussolini's greeting Hitler in Rome." May 7: "Japanese big guns check the Chinese in Shantung drive." May 13: "Japanese close in on Lung-Hai road [the Hankow part of what is now Wuhan]." May 14: "Franco forces advance against the loyalists [in Spain]." May 15: "Mussolini says Fascists will fight together to end if democracies make war."

On May 14, the *Greeley Tribune*, of which Walter was an avid reader, reported, "Italy's dictator, Il Duce, 'raps' the anti-Fascist attitude of the U. S." The *Washburn Leader*, which Margaret read, reported on May 17, "Junior-Senior Banquet and Prom was a 'Big Success.'"

The May 7–May 17 Letters

Each letter begins with the date it was written. They are ordered by the date they were postmarked. When multiple letters were sent in the same envelope, they are further ordered by the date they were written. The postmark date, time, and occasionally location follows the date that the first letter in the envelope was written.

May 8 [MAY 9, 7:00 A.M.]
Dear Walter,

"Let not your heart be troubled: believe in God, believe also in me. In my Father's house are many mansions; if it were not so, I would have told you; for I go to prepare a place for you." St. John 14:1,2. . . .

Walter, I am so happy that we have a home both an earthly as well as heavenly. God has always supplied us with our necessities and now He has also given me a promise of an earthly home with one who is a Christian provider. Really, I am grateful and glad. I just know that you will always be helpful, loving, and considerate of me. May God show me how to return thanks for it all . . .

This evening at supper the teachers asked me if sometime while they are travelling in Colorado if they could come to visit us. They surely love to talk about what farm-life is really like. This evening they said for me to be very careful when I am out gathering eggs so that I don't reach in the nest and touch a snake. There aren't very many snakes on your place usually, are there? Helen said one time her father was riding horseback and all of a sudden a large bull snake fell from a large branch of a tree right in front of the horse. The horse reared up on his hind legs and almost shook her father off the horse. I believe you said you aren't afraid of snakes, maybe I can get over my fright. Is there such a snake as a hook-snake that hooks itself up into a circle and rolls like a wheel? They told me about that but I believe that they use their imaginations when they get good and started in teasing.

Do any tease you? Not a single day goes by but I do get teased. But it doesn't bother me very much because I am being teased about you.

Today I have worn my Turtle Lake dress for the first time. My yellow hat matches the yellow flowers and the print of the dress. Mr. DeLange said he liked my nice hat. The hat is three years old this summer. I had it cleaned just before going to Colorado last summer. Since my dress has colors of blue salmon, white, and yellow on a background of black it makes my complexion not so sallow. Marie said all I lacked now was a little lip-stick. I don't use it and I don't intend to.

Thanks for writing about your finances. I am glad that you have paid your debts and bought and paid for your seed and feed. I presume the feed is for your three horses. I'll write some other time about furniture. I want to wait with something for my next letter. This morning for part of my prayer service I sang the song "I Need Thee Every Hour" and substituted the pronoun we for me. It worked out very well. Shall we use that song at each of our Sunday morning mutual prayer hour? It does express so much and then we can easily memorize it. When we get together next summer we can say it together? That hymn could be our theme song. . . .

Walter, I do long to see you. I do long to sit in your lap. I long to lay my head on your strong shoulder. Will you forgive me if I don't talk much because it just seems I can't although I am an English teacher? May God bless our being together. . . .

Yesterday I didn't do all the schoolwork that I had planned to do so instead I must go to school very early. I packed away some dresses and articles that I will not need this spring. I pressed some summer dresses that I had packed in my trunk from last fall. I believe the pupils will like if I wear something different. It seems most of my shoes need to be repaired but, since there is no good shoe-mender here, I plan to put them away until I get to Minnesota. Going up this hill three times a day is hard on leather.

Now may God continue to bless and keep you for Him, me, your dear ones, and your Sunday School.

Yours in Christ, Margaret

Margaret had a simple question: Are there snakes on your place? But she phrased it as "There aren't very many snakes on your place usually, are there?" I enjoyed contemplating why she, an English teacher, doubled the number of words necessary to pose such a simple question.

May 9 [MAY 10, 10:30 A.M.]

My Dear Margaret:

Blest be the tie that binds

Our hearts in Christian love:

The fellowship of kindred minds
Is like to that above.

This is one of my favorite hymns. It says so much with so few words, and in such an easy, simple, and understandable manner. May that bond grow stronger in each of us as the days roll on.

Dad sang this song from his heart. He often seemed mindful of instances when singing the second verse: "We share our mutual woes/Our mutual burdens bear/And oft before each other flows/The sympathizing tear."

I am so glad that I am very busy. It makes the time fly quickly. If it did not, and were not so, I doubt that I could remain so far from you and yet be at rest and peace within. So there is a purpose in the way all things are working and have been working.

How are you today? I am very well, in spite of freezing a great deal last week while riding the plow. I am through plowing now and I am preparing corn and bean ground. Today has been beautiful and about 7 o'clock this evening we had a little thunder shower, which certainly did refresh everything. Do you often have thunder showers in N. D. and Minn.? We have quite severe ones at times. Perhaps you saw and heard them last summer when you were here. Are you afraid of lightening and thunder? I do not like them too well but I do not greatly fear them, altho I do not take any unnecessary chances. . . .

I miss your letters a great deal, but I guess I can stand it until school is out. I hope then to hear more often from you. Until then write only as you have time and strength. This coming Friday I am to conduct a study of Paul's letter to Philemon. Please pray that I may not do it for my own glory, but to the glory of the Father in Heaven. Also that I may make it so interesting that they will continue such a good practice. . . .

You asked once about Mr. Brown's drinking. Sometimes he goes without whiskey for quite a while. Then again he comes home, drunk; every night for a while. He drinks beer all the time. I doubt that he ever quits alcohol. I wish I knew a way to help him. I try to remember him to God as often as possible and He can do much of course. It would indeed be a miracle if he quit now after using it for so many years.

Mrs. Brown and Asa are reconciled to it. He is not at all violent regardless of how drunk he is. It is certainly a terrible habit and I thank God that I never started using alcohol.

Each day brings more flowers and leaves upon the plants until now; it is very beautiful.

The sandman's nearly got me, so I bid you goodnight.

May the bonds of Christian love ever draw us nearer to each other and to Christ. Walter

May 11 [MAY 12, 7:00 A.M.]

Dear Walter,

"The earth is Jehovah's, and the fullness thereof: The world and they that dwell therein." Ps. 24:1.

Good morning, Walter. This is such a wonderful, beautiful world that we are living in. The birds are singing in their soft twittering manner this morning. The sun is trying so hard to peep through the clouds that I just wish that I could reach up to help it along.

Yesterday the junior choir had its picnic. There were four girls on the eats committee and they certainly did get enough food. We had potato salad, pork and beans, buns and wieners, nectar, three kinds of cake and ice cream bars. I supplied the last item. We played a few games but not as many as they wanted to because I didn't want to be responsible for them for such a long time. As soon as we had gotten started on our homeward walk which was to be three-fourths of a mile, I think they realized it was best not to have played longer.

My fountain pen has been fixed and now I am using it. It seems like a friend. It cost 99 cents to have it repaired. They put on a new pen point. . . .

Walter, I am so glad that you do not believe in buying things on time. My folks have never done it and I don't think it right to use something that is not paid for. I am sure there is some good second hand furniture that we could buy and then if all goes well the day will come when we can buy some new furniture. Our home to be isn't so very large but that most of the rooms can be furnished.

Today I wore a yellow dress that I have worn for at least three summers. Miss Foster says she likes it better on me than all my dresses except my "Turtle Lake" dress. Mr. DeLange wanted to know if I hadn't just put on a new dress. It certainly is fun to have something that others think looks nice. . . .

Say, I found out the meaning of the term to "spring-tooth" the ground. I asked Helen and she gave me the answer. [A "spring-tooth" is a rake-like implement with curved tines for loosening soil that is pulled by tractors now.] She surely knows a lot about farming. . . .

Thank you for the lilac blossom. It was still fragrant. The lilacs here have not blossomed yet. I do love to smell that flower. To think that Browns have so many bushes. The perfume which you asked about has all been used. I like the smell of perfume and I am glad that you do. John [a brother of Margaret] says he does and he especially likes when Ruby puts some on. He like you does not like much but "just a hint" as you expressed it. . . .

Walter, I am very glad that your work seems to be more enjoyable and easier this year. <u>I don't understand in what way I have helped you but I hope that I might always inspire you in some way</u>. God can give me what I need to inspire you always. It is going to be fun to have you help me wash the dishes and to be with you each evening. May God ever be our guide. If He is for us who can be against us? "For I know whate'er befalls us Jesus doeth all things well."

In Christ, Margaret

My board is 60¢ a day. Isn't that reasonable? The food is very good. It's going to be fun to plan menus for us so that we can get the necessary vitamins.

I'm glad your banker has so much faith in you. Did you finally sell your stored potatoes? I hope that your work will continue to prosper. I think it is fine that you are thru with planting and ready for irrigating. I hope you won't be out working with the water when it lightnings. I need you, so please stay away from the water at that time; won't you?

May 12 [MAY 12, 4:30 P.M.]

My Dear Margaret:

"And the night following the Lord stood by him, and said, Be of good cheer; for as thou hast testified concerning me at Jerusalem, so must thou bear witness also at Rome." Acts, 23:11

<u>I have often wondered at Paul's calm unconcern, in all persecution and affliction</u>. He some time before this had made plans to go to Rome. Now he was a prisoner and hard set upon, by his enemies. Later he was shipwrecked upon an island, snake bitten and many other obstacles came in his way. Yet he never wavered in his purpose to go to Rome, nor lost faith in God's leading.

Why? I think it is because of Gods promise and his nearness. And we too have a promise. Matt, 23:20 "and lo I am with you always even unto the end of the world." Paul only worked and worked. He left the worrying to God. Yet he never shirked any task. He never laid down and quit, with the thot that God would provide. Paul never expected to receive something for nothing and we too have to work and struggle.

. . .

Thank you for the letters I received. One on Monday and one on Wed. [May 11] I will answer some parts that should be directly answered. First we'll discuss snakes some. Tell those who tease you, that 99 out of every 100 snake stories are exaggerated to the point where they cannot be believed. Just plain baloney! One could be bitten by any snake found in the U. S. and not die if he cared for it at all properly. I think most deaths from snakes are the result of fright. There are not many snakes in the irrigated country of Colorado. I have never seen one in the chicken house. Besides the chickens will or would tell you. They would raise an awful racket if such an intruder came in the nest.

I experienced a chicken racket one midnight when my brothers and I headed for the henhouse with our .22 rifle and a flashlight. The hens were roosting on two-by-two rods placed parallel roughly a foot above a broad platform that stood a couple of feet off the ground. We lay on the ground in the doorway and aimed our flashlight at a hole in the dirt underneath the far side of the roost.

All was quiet for a quarter of an hour except for the occasional clucking of the sleeping hens. Finally, a rat's nose appeared, twitching with the smell of grain in the feeder. It edged cautiously out of the hole, possibly because of the unfamiliar beam of the light. When its head and chest were exposed, Paul squeezed the trigger. A deafening crack set off an explosion of squawking chickens jumping off their roosts, filling the air with dust and feathers.

That was enough for one night. We never did it again.

I am glad you don't use rouge or lipstick. When you [come] out here, you'll spend enough time outside to get the most wonderful complexion and it won't clog the sink when you wash your face either.

I do not know whether I can or will do much singing altho I surely do love the song "I Need Thee Every Hour," I hum, whistle and half sing a good part of my time. Perhaps that is no good for I never completely learn any song and often I do not know the name of the song or tune I have in mind. I believe you can help me and I hope I will be a good pupil. I shall try. I shall even try to memorize this song. . . .

This is an apt description of Dad's musical expression when he was totally enjoying his life. His songs seemed to be coming out of something buried in his soul and occasionally surfaced in a hum, whistle, a mouthing, and maybe a phrase or two.

You surely shall sit in my lap and your head shall lean upon my shoulder. One thing I assure you, that had you been a chatterbox type of person, I would never have come to Minn., not even tho you have many fine qualities. I shall surely look forward to evenings of quiet and rest with the girl I love more than any person in the world. I hope that when we talk, we can and will study and read together, so as to be able to talk intelligently upon subjects, other than people. I certainly do not care for any kind of gossip and I know that you don't either.

Now I know that the problem of whether or not to teach is bothering you a great deal. I wish I had never mentioned money at all. But it is necessary in our lives as we live today and the only medium whereby we can judge the value of anything. Perhaps God has not provided; what I thot was enough money, because he wants to show me that it is nothing to build upon[.] Certainly if we have love, the true love of Christ in us, toward each other, that nothing else matters and all else will be provided by the One who watches over us at all times.

God has indeed blest both of us, even since we met and in the light of this, I am bold enough to ask you to share with me what I have. I have only one reason against setting the date sometime in Nov.; that is this, should my work and time and also yours, allow that we could be married in Sept instead, it would mean that we could be together two months sooner and also not have such a rush of setting things in order just before Christmas. Margaret, I long so much for the day when you will be near me always. You mention difficulty in expressing yourself in this matter and I have the same trouble, but until I can take you in my arms and tell you how much I love you, these words must do. I love you and pray for you. May God bless us, prepare us as helpmates to one another, humble and mellow us so that our fellowmen and little children will love to be in our home. I hope God will give us children and then make us the kind of parents he would have us be.

In Christian fellowship, Walter

P.S. I just received a card from your mother. . . .

Just found, I am out of stationery envelopes. Please excuse this one.

The lived truths of religion are of a different nature than the factual truths of science, the felt truths of art, or the reasoned truths of philosophy. There are no hypocrites in science or art or philosophy. On rare occasions, a scientist may fudge the data, an artist may fail to hold interest, a philosopher may reason inconsistently. In such cases, he or she quickly loses the respect of others in the field but is not labeled a hypocrite. Not so with religious leaders who are found to live truths strikingly different from those they expound.

Paul's letters would come to define much of the Christian understanding of the life of Jesus after his resurrection. In his first letter to Margaret (*EG*, 11), Walter writes, "I like to read the letters that Paul wrote better than any I have ever read." Here Walter wonders at Paul's "calm unconcern, in all persecution" and his never losing "faith in God's leading."

Good for Paul—but also for Walter for not losing faith in God's leading as a result of coming to see life as Paul did. After rhetorically asking why Paul did not lose faith, Walter answers his own question and notes that he too has been assured the presence and promise of Jesus that sustained Paul.

Walter also knows that others will evaluate his and Margaret's witnessing to that presence and promise. In his March 2 letter (*QGW*, 148), he wrote,

> As you walk daily before your pupils and
> fellow teachers, your life is doing a personal
> work . . . that you are unaware of. You may be

sure that a pleasing, smiling greeting does not go unnoticed, nor any kind acts, even tho very, very small.

May 13 [MAY 13, 3:00 P.M.]

    My dear Walter,

    "Grace to you and peace from Jesus Christ."

    "Our times are in thy hands, O Lord, we wish them there. Our life, our friends, our souls, we leave entirely to Thy care." . . .

    Your last letter just made me long to write to you this morning even though it is as early as 4:45. I haven't had my pen to write with and I dislike to borrow someone else's pen because pens are easily bent in the point just by the different pressure exerted on them. I typed a page to you and then it just hurt inside of me because I have so often read "Do not send a typewritten letter to a friend" so I finally stopped writing and put the page into the waste basket after having torn it to pieces. Now, Walter, I am going to send you some letters even though they will sometimes have to be very short. I do love to write and oh, how I do love to get your letters. I get such a longing to see and be with [you] that a letter helps me to wait until the time comes and it makes me feel that you are near when I read it. May God bless our letter-writing to one another. . . .

My tests haven't been written yet. I just must get at them. We have to have Mr. Thorson check them before we can start cutting the stencils. This work of writing finals is no easy task and that of reviewing is just as difficult. The pupils are usually very uninterested in it because they feel just a glance over the material is sufficient but when they come to class they can't quite conceive why they have to know it better. I hope to give them a little quiz each day but that too takes extra time. Each day this week I have been up to school before seven o'clock to get work done.

I wish it weren't so cold for you out in the field while you are plowing. The rains which you have been having no doubt accounts for some of the chilliness. I wish that I could bring you some good hot coffee to warm you up.

Yesterday two of the teachers brought me about five pamphlets. I can't recall what the titles were exactly but I shall attempt: How to make the food dollar stretch, How to stretch the clothes' dollar, Systematic Consumer's Buying, How to Marry on a Small Income. If you were only here we could get some interesting ideas. I plan to read them myself. I wish that I could keep them but they belong to Helen and she needs them for her class reference material. . . .

Now I must sleep for a half an hour or I won't be able to do justice to my teaching. I long to see you and I surely feel sorry that my letter writing has had to suffer. I do love to hear from you, Walter my husband to be, God willing.

Good morning! It did feel good to sleep. Now Helen and I are waiting for our breakfast. It is going to taste good. . . .

I have just had dinner. The dust is blowing so that it is almost dark enough to have the lights on in the house. Someone said that the ditches in some places not so far from here are just filled with loose soil so that the road and the ditches are even. Isn't that sad? It doesn't seem to bother me much except that sometimes I feel like I am eating sand.

Won't it be fun to pick out the furniture? I was just wondering if I happened to have any furniture that I could bring with me. I don't believe I have. I paid for some of our furniture but the folks need all of it. Is the range pretty good that is in the house? That's right you eat your meals at home so you of course haven't tested it out yet. I hope that we can have it nice and clean all the time. I think a clean range adds much to a kitchen. The other day the Eng. III class gave a report of different authors. One poet's name was Morris. He had said "I will put nothing in my home except that which will be useful and will beautify it." I thought about our home and all the bright future. I surely was glad to hear the statement. . . .

Walter, you surely gave me a beautiful ring. I do love to wear it, but it makes me long to see you every time I look at the ring.

Tonight you are to conduct a Bible study. I would enjoy being in the audience. I'll remember you in prayer. It seems so wonderful to me that you like to give talks because I get so terribly worried and have to work a long time before I am able to speak before an audience to give a talk. It isn't hard for me to say a few words informally to a group though.

May you ever be in "Your holy Father's keeping."

Yours in Christ, Margaret

In separate envelopes, Walter sends Margaret a birthday card and this short (two-page) letter with the same dating and Greeley posting. It is followed the next day by another letter describing a conversation he had with a friend wanting to share a spiritual conversion.

May 15 [MAY 15, 3:30 P.M.]

My Dear Margaret,

"Grace and peace"

Indeed I do wish for you these precious gifts, far more precious even upon a birthday than any earthly gift.

This letter will be short and it is late. I beg forgiveness.

It is Sunday morning and I must soon go home for breakfast and then to Sunday School. I have to leave at 9 or immediately after.

I met you and was truly at peace a few moments ago during our mutual prayer hour. It is almost as good as seeing you. But even so, you cannot know how I miss you.

This will arrive late! Sometimes it takes your letters 3 or 4 days to get here. Does it take mine so long?

It will carry a little gift to you. The dearest person in all the world to me. I shopped and thot and thot and finally decided to send you money. Use it as you wish, for something you want or need in any way. It is a small gift and no measure of what you mean to me. There is no such measure regardless of how large. . . .

I thank god for you and thru Christ. I am with you. May he keep you in the way.

With much love, Walter

May 16 [MAY 17, 4:30 P.M.]

My Dearest Margaret:

God certainly moves, works and does in a deep and mysterious way.

I am so happy tonight I shall review this evening for you.

7:00 O'clock: thru eating supper

At 7:30, Leonard Molander called. I promised I would go to see him. (He is a good friend, belongs to the Mission Church in Eaton. We have gone fishing together many times.)

I had planned to visit him as soon as possible anyway. (I don't visit as I should.) All this; in spite of the fact that I had planned to come to Browns early in order to write to you.

Before I go further I must tell you that I had heard indirectly that Leonard had been converted at a Revival meeting in Greeley. I thot I might help him as a friend so I wanted to see him.

It was his birthday (same as yours) so his parents had a few friends mostly older people there to spend the evening.

9:30 had coffee & ice cream, ready to go home.

9:40 Leonard asked me to stay as he wanted to talk to me.

My spiritual growth has been slow extending over many years, as I believe yours has hasn't it?

<u>It is hard for us to understand what it means to receive a sudden vision or light, that Paul surely did.</u>

I believe Leonard (who is 25) is very sincere, but I know it will be hard for him (his old friends will not accept him now and <u>he felt I was the only one he could talk to and he must talk some</u>).

It also helped me much to know, that God works always and that the gospel still is the power unto life eternal.

Margaret please pray and ask God to make me a worthy friend of Leonard's as well as of yourself. Ask God that I may not glory in this confidence but that I strive and be willing to help Leonard in order to glorify him who died for us.

I am not worthy of your confidence or his. I am not 1/10 as good as either of you think I am. I neglect to visit my friends and S. S. [Sunday school] class. I am vain and glory in my own doings and strength. I am not as humble as I should be.

<u>I was so glad that I could testify that God answers prayers.</u>

<u>I told him how I had always prayed for a Christian wife and home and how God had answered that prayer, by sending you.</u>

Tonight you mean so much to me. To think that I need to seek no one to speak to but can come to you and discuss anything and everything. I haven't thanked God enough for you.

I love you so much.

Please do not worry about the letter writing. You are doing fine. I wish you did not have to work so hard. I can wait until school is out. A short letter now and then will do. I wish I could help you. I will pray that your strength will be equal to your tasks.

If God wills it, and it shall be my prayer, you will teach your last this year. I want you here to help me and besides I want to take care of you and help you.

It will be fun to shop for the things we need and I am sure that the necessary provisions will be made.

It is 12:30 and I must sleep some. I am planting beans now. With a good day I will be almost done tomorrow. We have wonderful prospects now.

Our Father in Heaven will guide, lead and keep you always.

God is indeed good!

In Christian love, Walter

P. S. Please pray also for this young man, for he will experience some hard days.

Walter is describing a highly personal conversation with Leonard, a good friend and fishing partner. Walter no doubt reflected on that discussion during his ten-minute drive home from Eaton, where Leonard attended church. He must have written this letter as soon as he got home, as he signed it shortly after midnight.

On the way over to Leonard's birthday party, Walter might have been contemplating the article "Lutheranism or Revivalism," mentioned in his April 19 letter (*QGW*, 372). Dad would have sided with Lutheranism, but the article would have left him wondering if it wasn't a lack of love that leaves us perceiving only the faults in others.

In the conversation, Walter emphasized some of the consequences of Leonard's new life as a Christian by sharing a few of his own spiritual experiences: seeing God at work in his life; answers to prayer, especially his prayer for a Christian wife; and an assuring awareness of eternal life. He apparently pointed out that Leonard could lose some "old friends." In parting, he might have prayed that Leonard would find new friends and that their mutual friendship would deepen.

Although Walter didn't spell out the details of Leonard's conversion, he was certainly thinking about them as he drove home.

In his March 2 letter (*QGW*, 148), he had found "a good answer to the question [from John 3:3]: Do we have to be born anew, and what does it mean?" His answer came while reading that the spirit of Jehovah came mightily upon Saul and turned him into another man (1 Sam. 10:6) and gave him a new heart (1 Sam. 10:9).

More than once, Dad told me how in high school he decided to stop spending time in pool halls as a consequence of his taking seriously his emerging Christian understandings. Here, he emphasizes that his "spiritual growth had been slow extending over many years," a "Lutheranism" he shares with Margaret. Thus, he is finding it "hard to understand what it means to receive a sudden vision." Yet he wants Margaret to know that Leonard is sincere. He even suggests a possibly surprising depth to Leonard's spiritual transformation by linking it to Paul's dramatic conversion. Here it is in Paul's own words:

I want you to know, brothers and sisters, that the gospel I preached is not of human origin. I did not receive it from any man, nor was I taught it; rather, I received it by revelation from Jesus Christ.

For you have heard of my previous way of life in Judaism, how intensely I persecuted the church of God and tried to destroy it. I was advancing in Judaism beyond many of my own age among my people and was extremely zealous for the traditions of my fathers. But when God, who set me apart from my mother's womb and called me by his grace, was pleased to reveal his Son in me so that I might preach him among the Gentiles, my immediate response was not to consult any human being. I did not go up to Jerusalem to see those who were apostles before I was, but I went into Arabia. Later I returned to Damascus.

Then after three years, I went up to Jerusalem to get acquainted with Cephas [Peter] and stayed with him fifteen days. I saw none of the other apostles—only James, the Lord's brother. I assure you before God that what I am writing is no lie.

Gal. 1:11–20, New International Version (NIV)

Differences in political and religious identities and allegiances can be divisive. Leonard could be viewed as a "revivalist" and Walter as a "Lutheranist." But that is not how they saw each other. Instead, each saw a friend sharing different spiritual experiences that were opening up their worlds. Walter came away from their conversation with a renewed joy in seeing God at work in a "deep and mysterious way." I expect Leonard did as well.

May 17 [MAY 17, 3:00 P.M.]
> Dear Walter,
> "Where He may lead, I'll follow
>   My trust in Him repose
> And ev'ry step of my onward way
>   I'll sing He knows; He knows."
>
> This is a song that I almost know from memory. I am so glad that He knows every step of our onward way. It feels so different every time I think that He knows and directs.
>
> Last Saturday I received your letter telling of how Paul withstood all but was firm to the end in working for His Savior. His faith grew and deepened in Christ with each trial. Paul is a strong example to have before one when one has a trial.

The subject of snakes has not been brought up since I wrote to you but now I have my answer ready. <u>I presume that fright plays an important part in the snake's deadly poison. If people knew how to properly care for the wound then all would be different.</u>

Mom eventually lost her fear of snakes and entered fully into life on the farm. We had a little brood house where she enjoyed feeding the chicks as they ran around her feet softly peeping. She enjoyed watching them grow, and once they were grown, gathering the eggs. It was a special treat to watch her catch a rooster for the evening dinner. She would grab the chicken hook—a long, stiff wire bent into a V at the end. It took only one or two thrusts before she hauled in a flapping rooster after catching its foot in the hook. She would tuck the rooster under her arm, smoothing its feathers as she marched to the chopping block. With its wings and legs in one hand and a hatchet in the other she soon parted its head and body.

After the bird had been hung by its feet for about an hour, we would dunk it in scalding water so it would be easy to defeather. Mom would then gut it and cut it into serving-size pieces. If we were around, she enjoyed opening the crop and gizzard so we could see what it had been eating and the little rocks it had ingested to break up the grain stored in its crop. A few hours later, she would serve up the anticipated chicken dinner.

Sunday was a big day for me. I rode to Wilton with Rev. Johnsons in the afternoon. I played the organ for the services. Rev. Johnson said that he wanted to express his and the congregation's appreciation for my playing and singing. Then he added that I was not returning to Washburn next year but was planning to build a home of my own with someone in Colorado. Now you see, "we were cried last Sunday" as a Tennessean would word it.

    Walter, I am so happy about it—your wanting me. I think it will be so much fun to be with you. "We'll build a sweet little nest out there in the west," won't we? It surely will be fun to plan. May God give us wisdom to know what His will is in all our planning.

    Sunday night we had Luther League out in the country. I had had a headache so I called them[,] the program committee[,] at suppertime that since I had not been feeling well since morning I thought best to stay at home. Then after supper my headache left me so I went to the church where we were to gather for transportation. Were they surprised to see me? I sang the song "Living for Jesus" from the book that you sent me so in a way you were present too. After the program they wanted to see my ring and to know about you. I told them and the ring surely received a lot of praise.

    Last night Mrs. Thorson invited me to come to visit with her mother as I had told her I wanted to hear more about her travels in Europe. Not very long after here came all the teachers singing "Here comes the bride."

Was I surprised? We spent the time talking; then, lunch was served, which was maple nut ice cream and a birthday cake with the candles spelling out the initial "W." Then they turned the cake around and they spelled or formed the initial "M."

They asked me a lot of questions about you and gave me much advice about what to do on a farm and so forth. Mr. Thorson presented me with a booklet that contained little expressions from each teacher and Mrs. Thorson.

Then Mr. Thorson took me downstairs to a bedroom and there on the bed was a blanket all dressed up as a baby. I carried it up, and when I got up, they told me to shake the baby. Then I heard some music from inside; I reached in my hand and pulled out a music baby rattler.

The blanket is all-wool, and it was decorated with wooden spoons, measuring spoons, chore boy, and vegetable brush. I certainly am happy about it all. Then after I came home there were two packages on the dresser from a junior sophomore girl who works for Jefferis. They contained salt and pepper shakers and a knife holder box. You'll get to see all these things in action sometime, God willing. . . .

I am going to write in for a teacher's life certificate. I qualify so I believe it is best that I apply for one.

Now it is time for me to go to school. The recital at which I am to play a flute solo is to be held this Thursday. I haven't practiced so tonight after school I shall do so. I hope we can get a piano but if we can't we can get along anyway. I would like to teach you to play hymns. Wouldn't you like to?

Now may God bless and keep you. It will be fun to come to our home together either in November or September.

Lovingly, yours in Christ, Margaret

Although Margaret's writing "in a way you were present too" suggests her conscious awareness of Walter's spirit, it belies the extent to which his spirit is materially present and alive in her. She is its primary voice in Washburn. News gets around because his is an infectious spirit. Her friends and colleagues would love to see Walter—but he is in Colorado, most likely moving about either his parents' farm or his rented farm near Eaton. So they turn to Margaret. She writes that those at the Luther League want to know about the person who gave her the ring. A paragraph later, we learn that the Thorsons want to know about Walter, as do her fellow teachers and the girl "who works for Jefferis."

Margaret enjoys Walter's spirit. That is why she wants to talk about him. She can because his spirit somehow materially and functionally resides in her brain. Were that not the case, she—like the others in the Luther League—would also wonder about the significance of the ring on her finger that "received a lot of praise."

That ring is the only material evidence of Walter's spirit the Luther League members can see and touch that is physically distinct from Margaret's personal presence. She could point to the materiality of the expressions nourishing Walter's spirit within her—his letters—but she is filing them away to possibly revisit another time. Thus, the only aspects of Walter's spirit that are now entering and residing in the minds of her colleagues are first being filtered, modified, and clothed in her own words before being passed on to them. Judging from their letters, she and Walter are reasonably truthful and expressive. That said, everything of Walter's spirit that is embodied in his letters and feeding the growth of his spirit in Margaret's mind is screened, undressed, modified, and clothed again before being shared through her expressions with others.

We are decades removed from what of Walter's spirit Margaret passed on to her colleagues. All we have is her brief descriptions of what she told them. Consequently, much more of Walter's spirit probably resides in us than in them. In these letters, we have the very expressions that fostered the growth of Walter's spirit in Margaret's mind.

>May 17 [MAY 18, 3:00 P.M.]
>Dear Walter,
>"I will bless Jehovah at all times:
>His praise shall continually be in my mouth."
>Ps. 24:1–2

Thank you for the letter, greeting, and gift. Walter, I surely thank you. I am sorry you could not find something but I will keep it until you come to Minnesota and then we can pick something out together; won't that be fun? There are so many things that I have thought of that we could buy. . . .

This evening I have written an application for a life certificate in teaching. I hope that I receive the blanks sometime this week because it will save me much time. There are so many ways in which I realize now I could have taught better that I wish I had thought of before.

You mentioned that some of my letters do not reach me [you] until about three or four days after I have mailed them. The letters which I receive from you often take three days but there are times when I receive the letter only two days later. The letter post marked 3:30 p.m. May 15, Greeley reached me this afternoon at 3:45 p.m. I didn't mean to repeat. That means just two days for the letter to come way up here. . . .

Don't be surprised if you get an air mail letter from me. This is the first time a mail plane is stopping here. Mr. Nelson, the post-master, asked me if I didn't think I had better send an air mail letter to Colorado. It didn't take me long to decide but the letter is not anything urgent. I wonder how long it will take for it to reach you.

Tomorrow I plan to pack down our all wool blanket. This summer I hope to sew a quilt. We've got to keep warm. I want to embroider a little design on a pair of pillow slips that I bought after Christmas.

Mrs. Jefferis was up this evening. [Margaret and her friend Helen rented and shared an upstairs bedroom from Mrs. Jefferis for the nine months they taught at Washburn.] She told me last night about what hard financial difficulties they were in for about four years of their first years of marriage. She said she was glad that they hadn't waited to be married. Walter, <u>I haven't told her anything about finances except last spring</u> when you wrote that you were coming. I told her of your plans and then I received word that you couldn't, so I had to tell her that your plans didn't materialize because your stock hadn't sold as you had anticipated. I think that Mrs. Jefferis sees how the farmers are faring up here and so I told her how your farm was equipped and the house wired and had running water. She was surprised.

Have you had some pictures taken of the lilacs? I wonder if someone would take a picture of you soon, just a snap shot. I would like to have one.

It is nearing 12: o'clock so I had better get to sleep or I won't do justice to my teaching. We're planning to get to bed earlier; aren't we?

Maria Olson sent me as a nice birthday letter and a package containing some wearing apparel. I can't understand why she should send me a gift like that. She wants to be a missionary but her folks discourage her and will not help her to get a chance for further education.

Thank you again for the birthday gift. You and I will have fun buying something. I want something lasting, useful, and beautiful. I am glad that you do not mind that I am older than you in years. Oh, it isn't years; it isn't quite a year. God has been kind to me all these years. It seems this past year He has given me so much that I don't know how to thank Him. The future appears bright. It remains in His care. "As thy days, so shall thy strength be," he has said.

Much Christian love to you, Walter,

Margaret

Good morning! I am just about ready to go to breakfast but I decided to add a few lines to this letter. Since I have no convenient place to write the writing part does not go very smoothly. . . .

Won't it be fun to have your folks over for meals? I hope our home will be such a one that they will always feel welcome. I do so want to fit in your family.

Time passes along. There are only six more teaching days; then away I go sailing to Minn. eager to spend the summer and to prepare for your coming. May God draw us nearer to Him and to one another.

Yours in Christ,

Margaret

A Glimpse Back

As in the preceding two volumes, I will be including at the end of each chapter expressions raising self-exploratory issues worth discussing. Although each expression stands on its own, should you wish to know more of its context, it is underlined in the letter written on the indicated date.

Because of the importance of these expressions to the love song that my parents were singing, they will be important to the songs that I will be singing. As suggested in the preface, our basic concerns for life, disclosure, worthiness, relationships, brokenness, and death give rise to our guiding spirits. Consequently, these selected expressions are summarized in Appendices 1 and 2 in ways that bear on these universal concerns. A sense of their distinctiveness and the spirits they evoke can be gained simply by glancing through these Glimpse Back expressions.

May 8. When envisioning "an earthly as well as heavenly" home, Margaret mentions God's providing the necessities, the promise of living with a Christian provider, her trust that Walter will be "helpful, loving, and considerate," and her thankfulness. How might these understandings shape her life with Walter?

May 11. In his March 2 letter (*QGW*, 149), Walter writes, "If just a letter from you lifts my spirits . . . think what it will mean to me to be able to come to you at any time." Here she "hopes that I might always inspire you [Walter] in some way." She then lists some things involved in her realizing that hope: a gift, a chore, an evening, a guide, and an understanding. How do you see each of these things contributing to the realization of her hope?

May 12. Walter is seeking for himself the source of "Paul's calm unconcern." He finds it in the voice of Jesus: "Lo I am with you always" [Matt. 28:20]. In what ways is the material presence of Jesus's spirit in Walter's life in Colorado the same as or different from the material presence of Walter's spirit in Margaret's life in North Dakota?

May 16. Concerning Leonard's conversion, Walter writes, "He felt I was the only one he could talk to and he must talk some." Have you encountered life-changing experiences that you wanted to share with others but were uncomfortable expressing? What lies at the heart of that discomfort? How can it be overcome?

May 17, first letter. In her May 8 letter, Margaret was concerned when her colleagues warned her about potentially touching a snake when gathering eggs on the farm. In this later correspondence, her fear of snakes is seemingly resolved by her seeing fright as having "an important part in the snake's deadly poison" and knowing "how to properly care for the wound." While some of our fear is sourced in experience, some comes from the opinions of others. How do you see the different roles these two sources played in the creation and the resolution of her fear of snakes?

May 17, second letter. Surprises arise when we are not fully aware of what is going on. Margaret is surprised at the awareness Mrs. Jefferis has of her and Walter's hesitations in setting a wedding date because of monetary concerns. How might Mrs. Jefferis's love and interest in others have been involved?

The Night of the Little Self

Walter devotes his May 16 letter to a spiritually transformative moment in the life of Leonard, a close friend, and relates it to the "sudden vision and light" that transformed the life of Paul, who later wrote the bulk of the epistles in the New Testament. I can't speak knowingly of their transformative experiences. I can speak knowingly of mine.

It was sourced in a quest that may have begun when Mom offered me five dollars if I would read through the Bible. I started reading a chapter or two each night. On paying me a couple of years later, she repeated the offer, so by my mid-teens, I had read the entirety of the Bible twice.

Possibly because I never questioned the truth of the Bible, its words profoundly affected me. The Hebrew Bible (the Old Testament) left me with a sense of the awesomeness of God. The New Testament left me with a deep appreciation of the love and sacrifice of Jesus. Both testaments created perplexing questions that would resurface from time to time throughout my college years. The presence of God . . . miracles . . . revelations, what were they? They seemed to be tied in to the nature of that part of me that perceived them—my mind. That too was a question.

Three related conflicts frequently resurfaced in one form or another in college during discussions at meetings of the Lutheran Student Association:

- Miracle after miracle in the Bible defied the assumptions of science.
- The limited time domain of the creation accounts in the Bible conflicted with the astronomical time frames of the creation accounts in science.
- Truth in science was founded on objective observation and testable theory; biblical truth seemed to be founded on historical accounts of personal revelations from God.

For the time being, I dismissed the first conflict as being essentially unresolvable. The only documentation of the miracles with which I was familiar was described in the Bible, and that was the very document under question. It seemed you either accepted it as written or you didn't.

I also tabled the second and third conflicts. Who's to say God could not create the world through the processes of evolution? Moreover, answers either way had little bearing on my sense of God's pervasive awareness of my thoughts and actions. That left me with more personally relevant questions: What is God? What is mind? And how are the two related?

My questions simmered without clear answers or any sense of headway throughout college and into my first year in a PhD program in experimental statistics—until my older brother, Paul, unexpectedly called, saying, "Now I know what God is. God is truth, pure and simple truth." He came to that realization while reading the book *The Origins and Growth of Biology* as part of his PhD studies in animal behavior. It changed his life and energized his studies. Although the book also expanded my interest in the nature and transparency of scientific truth, it didn't engage my awareness of God as a presence in my life.

A few years later, I ran across the book *Mind, Self, and Society* in Paul's office. It contained the lecture notes of George Mead, known for his work in social psychology, as compiled by a few of his graduate students. The notes addressed notions related to my vague sense of the mind. Mead saw gestures as a means by which animals communicate with one another. A gesture could be anything—a crouch, a bristling, a growl, a scent, a wink, a fist, a failure to act, even disinterest. Every state, sound, and movement of an organism signaled something potentially perceived by another.

Mead emphasized that the meaning of a gesture depended on its perceiver. The song of a wren singing in a bush may mean one thing to its mate, another to a cat, and another to a poet. But suppose you turn around in a crowded theater and see flames. You might instinctively yell, "Fire!" In that case, your expression would mean roughly the same to all in the theater. The actual fire would spread in slow motion relative to the speed with which your gesture would light up the audience. Mead termed expressions that stand for roughly the same thing in the mind of the sender as they do in the mind of the perceiver "significant symbols."

I had always thought of my mind as something localized within my head. Now I was seeing it as somehow coextensive with the minds of others. I had resumed reading Mead's book in my upstairs office one evening, contemplating this new sense of the mind, when luminous fibers suddenly shot out from my midsection. Truth impressions and expressions of which I could catch only the general gist were streaming in and out at numbing rates. I was no longer my body or even my mind (when viewed as something confined to my brain). I was a node caught up in an incomprehensible web of influences through which all of us were somehow connected. I was enlivened by a vast and encompassing spiritual presence of which I was inescapably a part.

At first struck dumb, I began excitedly pacing the room, trying to contemplate what had happened. I, who had so long pondered what a revelation from God might be like, had just experienced one! A dispersive existence displaced my former sense of a materially localized one. Self-importance gave way to inherent self-worth. A material lifetime that waxed and waned in a few short years shrank to little more than a moment in a spiritual existence that somehow originated in the distant past and extended far into the future. The kingdom of God—its joys, its understandings, its relationships—sprang to life. My mind jumped back to an understanding Jesus had expressed. I was in God, and God was in me, and, by the same token, I was in my neighbor and my neighbor in me. I was caught up in an inclusive spirit of acceptance and freedom of which I would always be a part.

The web was so simple. Each expression was materially received, interpreted, and possibly reciprocated. At the same time, its dynamic was awesome and complex. The expressions were endless; the cumulative effect of even the simplest one, potentially profound.

In one sense, nothing changed in that revelatory moment: same material world, same body; no smarter, no new facts or skills. In another sense, everything changed: who I was, how I saw others; I had a whole new way of looking at the world and at what I was about. Walking up the stairs that evening, I was caught up in a competitive "big self" view of the world, proud of my accomplishments—PhD, rewarding job, happy marriage—but wondering what gestures had to do with my mind and my sense of the presence of God. Walking down later that evening, I was caught up in an inclusive "little self" view of the world in which my being and circumstances were incomprehensible and ongoing gifts of a universal and "eternal" web of relationships in which I was inextricably caught up and totally dependent upon. Life-changing experiences like that have to be expressed. But how?

I was not given the needed words and reasons, possibly because the experience was closer to my heart than my head. (Blaise Pascal, a seventeenth century mathematician, philosopher, and theologian, said, "The Heart has its reasons which reason knows nothing of" in regard to his mystical experience now called his Night of Fire.) I returned to the Bible for the words it had used to contrast the world of the inclusive little self with the world of the competitive big self. Some contrasts began to stand out: a unifying spirit versus a divisive one, a self-insistent spirit versus a selfish one, a truthful spirit versus a deceptive one, a humble spirit versus a superior one, a self-responsible spirit versus a faultfinding one, a generous spirit versus a greedy one, a serving spirit versus a dominating one, and an immortal spirit versus a mortal one. Because of the intuitive nature of these spirits, I knew similar contrasts would be found in the scriptures of other religions. Finding that to be the case (*QGW*, 115–121) only increased my need for the critical words and reasons. They finally came in augmenting these letters with my songs.

CHAPTER 2
I DON'T MEAN TO BE INDEPENDENT OF YOU

The following are some of the headlines from the *New York Times* contemporaneous to the letters that follow. May 20: "Suchow is occupied by Japanese"; "Chinese rearguard actions sharp"; and "Czechs put 400,000 troops near borders of the Reich." May 22: "Insurgents claim main road to sea. . . . Madrid shelled, 12 killed." May 23: "Organized Nazi terrorism goes on secretly in Vienna." May 25: "Paris asks U. S. diplomatic backing to prevent a German-Czech war." May 27: "Accord is reached on troops in Spain by all but Russia."

The *Greeley Tribune*'s headline for May 21, "Weld's May rainfall is far above average," was something Walter already knew and certainly would have welcomed.

This chapter's title, taken from one of Margaret's expressions in this section, belies the anticipation of the joy in married life Walter and Margaret will share in the next few letters. It more reflects the confidence they have in being themselves within their love for each other, a confidence similar to that of which I will shortly sing in "A Helper Fit for Me."

The May 17–May 25 Letters

May 17 [MAY 19, 4:30 P.M.]

Dear Margaret:

Jehovah is my shepherd; I shall not want. Psalm 23:1

Thank you for the letters. They are helpful and do help me. They encourage and inspire me and help me to carry on each day until the day comes when I can speak to you face to face. Maybe I won't have much to talk about at times and I will only want my darling, near to me for the comradeship which no one else can give. As I think of it and look back to when I first spoke to you alone, <u>I cannot remember a single time when it was hard to speak to you</u>. It has always been easy and natural.

You mentioned not long ago something about not caring to talk all the time. I am so glad of that for I am the same. It seems to me that we have much in common.

I certainly am not in want. Jehovah has blessed me far beyond my worth. . . .

I want to say now; that if it is easier for you to type letters, then type them. That will certainly be alright with me. I wish I might help you more than that. When we are united in marriage may we always strive to be a mutual help to one another and then our love will grow and not diminish with the years.

Only write as you have time. Tho they mean so much to me yet I love you much more than them, and as I know you are working very hard now, I shall try to get along without them. And they need not be long.

I am afraid you work too hard and worry too much. I wish I had you here. The girls tease you about the hard work but I do not believe it is that bad, and *besides if you were here I could help worry. At least I would try to do that, as well as encourage you in any way I could.*

It rained and hailed some about 10 o'clock today. I had to quit planting beans and run home, but I got soaked anyway and had to change clothes. By the time I got home and had the horses in the barn, it had quit raining, so after dinner I went out and planted beans all afternoon. I like to plant any crop. I will finish planting beans tomorrow if it don't rain. Then I will only have potatoes to plant yet (about the middle of June).

I will have 27 acres of hay, 17 acres of barley; 18 acres of beans and 18 acres of potatoes.

We will get water [for irrigation] anytime, now. (I suppose on Sunday. We often do and generally at night also.) Then I shall be busy for awhile. Next year I hope you will go with me some to hold the lantern but mostly for company.

Wed. evening 8:30 P.M.

"But if we walk in the light, as he is in the light, we have fellowship one with another, and the blood of Jesus his son cleanseth us from all sin."

My dearest Margaret, tho you are nearly 1000 miles away, yet I feel the power of the fellowship, which we have in

Christ. It does not seem that you are so far away. Many of our prayers are being answered I know. For one thing we are having the most perfect kind of spring. I finished planting beans, today noon. Then about 4 o'clock, it started raining, and we got enough rain to make the beans come well and quickly. Isn't that fortunate and nice?

I and my brothers always welcomed a half inch or better rain which meant no irrigating and fields too wet to work, thereby opening up the possibility for going fishing. We hurriedly did the morning chores. We didn't race though; Dad would have chided us for lazing around on the other days. But he encouraged an early start and only smiled when the chores were done expeditiously.

Margaret you need not worry about either buying or bringing any furniture. I am sure I shall be able to provide that. This money question need not worry us at all. I am not going to worry about it any more. Our needs will be adequately cared for. <u>I shall work and take every possible care of all my crops and then take what I receive</u>.

Mr. Brown has been pretty sober for quite awhile, but tonight he came home hardly able to walk or talk. I don't know how this will work out when you are here. He may bother us quite often. He does me, but if he bothers too much after we are married, we will get another place to farm, altho this is an especially good one with plenty of water as a rule.

. . .

Everything is green now and very beautiful. Browns yard is especially beautiful. I wish you could see it.

I must shave before I retire, or the animals will not recognize me. So I bid you goodnight.

May our Savior give you understanding, knowledge Peace and rest.

In Christian love. Walter

Walter had the post office stamp this airmail celebration before he addressed it.

I don't recall Dad ever saying much about using time except not to waste it. He liked to work. He could have worked harder, but I think it was one of the keys to his joy that he took roughly forty-five minutes for breakfast, an hour for lunch and an hour for supper and savored his food. In the evenings, he read the newspaper and magazines. Sunday mornings were spent in church; the afternoons were spent resting or doing things with the family.

My taking a sequence of three core courses in zoology my senior year led me to a similar view of using time. The first was histology. I struggled with the first test and was disappointed with the outcome. The handful of A's and quite a few B's starkly contrasted with my C. I had a B+ average going into the course and had hoped for A's in this core sequence. I bore down in my studies for the rest of that fall quarter, but to no avail. Half the students were in graduate school taking nine to twelve credit hours and funded by various assistantships and fellowships. I was taking eighteen credit hours and working twenty hours a week to cover my living expenses. I studied whenever I could but still ended up with a B.

That winter, I signed up for the second of the core courses, embryology. Ahead was the same scenario—competition with the same graduate students and the same eighteen-hour workload—except that I had had enough. I limited my study to two hours a day so I would have some free time to do whatever I wanted.

It worked. Instead of cramming for the first test, I read over my notes, wrote down any words I wanted to remember, and decided to let the grade fall where it may. I aced the test and, a few months later, the course.

Now whenever the realization hits me that I am working harder and enjoying it less, my mind goes back to that course in embryology. I slow down. I limit what I do to what is most essential and can be completed in time to enjoy the evening. It always seems to work. The joy comes back, and, strangely, I soon start feeling that I'm getting more done.

May 20 [MAY 20]

> My dear Walter,
>
> "Search me, O God, and know my heart:
> Try me and know my thoughts;
> And see if there be any wicked way in me.
> And lead me in the way everlasting." Ps. 139:23,24.
>
> This is Friday morning. The last regular school day that I am to teach this year, God willing. I can hardly realize it all. . . .
>
> After senior choir practice Mrs. Handy suggested that we go down to the basement to practice through the hymn which we sing down there every Sunday morning. We all went down and what did I see? The basement decorated and a long table set. It was a shower for me. I couldn't realize that it was for me so I asked if it was a surprise on Rev. Berg for his birthday. I don't recall exactly what they said but I learned that his birthday was to be the next day. After eating lunch they gave me a handkerchief shower. Walter, I have never in all my lifetime had so many beautiful handkerchiefs. I am not going to use them now. I want you to see them first because even though they are a lady's handkerchiefs you have part ownership. I have not mentioned that one handkerchief was a large dark red one on the greeting card were printed these words, "This for the bull." . . .

I am so happy about Leonard's conversion. To think that god saw fit to use you to strengthen him. One soul is of great value to God. May God continue to help you in winning souls for Christ and strengthen him in the faith. I have remembered him in prayer and I hope to continue to. God does truly move in a mysterious way. . . .

Now it is just beautiful outside. I wish we could take a walk together but we'll have to wait until you come to Minnesota. It will be fun to arise from sleep to know that you are near. Just think when the time comes when we shall awaken to see and know that the bridegroom Jesus is near. Yesterday I listened to a broadcast of organ and mixed chorus rendering the song, "The Lost Chord." It made me fathom to a little extent the beautiful music that we may have a part in in heaven. It just thrilled me and made me long for that blessed home. May we be ready when He calls and may we get "to see our Pilot face to face when we have crossed the bar" as Tennyson expresses himself.

Much love to you and may the grace and peace be yours, dear Walter, always.

Yours in Christ, Margaret

Margaret's celebration of National Air Mail Week cost her double the usual three cents.

The varied ways Walter addresses potential differences and aggravations in his next letter are natural backdrops for some of his ways that played out in my childhood.

> May 20 [MAY 21, 4:30 P.M.]
>
> My Dear Margaret:
>
> "Let all bitterness and wrath, and anger, and clamor, and railing, be put away from you, with all malice: and be ye kind one to another, tenderhearted, forgiving each other, even as God also in Christ forgave you." Ephesians 4:31,32.
>
> To be able to do all these is my desire and goal. This is one of the reasons that I am so glad that you are a Christian girl, for I know that you to[o], have this as your goal. I am not so much an idealist, as not so much to realize that I will fall far short of this wonderful ideal, or that you too cannot fulfill

it entirely. Nevertheless; to have this in common will be a great help to each of us. I am determined that <u>we shall take any little differences that may arise between us, to God</u>. We will do this, not separately only, but together and all this from the very first. I can safely say; that I am determined, because I know this to be your wish also. I could not do this, if you were not a Christian and you could not, if I were not. It is thru this fact, that I know how much you love me and long for the day when we may be together always. It is absolutely impossible for me to put any measure of my love upon this paper and I truly know, and honestly I have not one slight doubt, that your love is fully as great or greater. Seeing your concern for your fellowmen, a few times, makes me think that your love is greater than mine. May God increase my love, to make it worthy of the girl, that he made it possible for me to have.

I've no doubt they did bring many of their little differences to God, but if so, it was at night or in the privacy of their bedroom. Generally, their disagreements were ironed out openly in front of us children without bringing God into the argument. They voiced their frustrations. Their needs and reasons were argued clearly and forcefully.

Disagreements often started with Mom asking Dad to shop for groceries or run some needed errands when he was rushing off to get parts for a farm implement. He was busy. She was busy. If the disagreement heated up with no obvious end in sight, Mom would change its direction by saying, "Walter, you still love me, don't you?" Her expression was less a question than a desire to touch again the foundation of their relationship.

Voicing such a need in the midst of an emotional argument takes faith in that foundation. Dad must have always sensed that the argument was now cutting into something he deeply valued. He never uttered a heat-of-the-moment and possibly devastating "no." Instead, he would flounder around for a moment and then reply, "Margaret, of course, I love you." With that, the argument would end, and a mutually acceptable understanding would somehow be quickly reached.

> I am so glad for you; that you have such fine friends in Washburn. It certainly was nice of them to give you these nice presents and it sounded as though all were practical. I should like to see the presents and I hope to soon. I wish that I could have been with you when we were cried, as you said. I have said nothing here, altho I can sense that it is understood in fact I was congratulated once.
>
> Though I am bashful, I am very proud of you, if I may use that expression, I know no better. I would gladly stand with you to be cried. May God make me more diligent in praying for us, praying that we may be given the necessary qualities to heed Paul's admonition. Then will that little home "in the West," truly be a place where people will love to be.
>
> . . .

It is certainly fresh and beautiful after these rains. When the sun shone after a shower yesterday, I longed for you, that you might have seen it with me. Again I say that I have never seen such a beautiful spring. This cloudy weather also makes some very beautiful sunsets.

If I have it right, you will be thru with your school the 27th of May. It won't be long until I must address my letter to Clarissa. You will be happy to be home again won't you, even tho you have so many friendly people to be with, in the place where you are now. I wish I could come in to Washburn about Fri. noon and take you home to Minn. But I shouldn't think on it, for it only makes matters worse, because I see no possible chance of doing so. . . .

I am told that the water will come Monday. It is hard work but I am glad to see it come. For without water this country is useless, for farming. First I will irrigate the hay, then the grain and if I have enough early water (that is, direct from the river. Later on we get it from the reservoirs) I will irrigate beans also and maybe enough for the hay and grain again.

Mr. Brown is on another drunk again. Boy, do I hate whiskey. I wish I was authorized to blow up every distillery in the U.S. I don't see why they don't take one of the cyanides and get it over with in a hurry. That is what they use in the gas chambers for executing criminals.

Dad was seldom angry and did not carry his anger with him or bury it inside, yet it could be aroused. One time in particular sticks in my memory, not because of what I did, but because of what he did.

Dad's disciplinary judgments were usually deliberate and consistent with the 13:24 proverb, "He who spares the rod hates his son, but he who loves him is diligent to discipline him." I remember my uncle Bruce tolerating some back talk from one of my cousins in a manner Dad would never have tolerated, which may have led to Dad's occasionally remarking, "It's just not right that Bruce does not discipline his children." When I became a parent, I had to reconcile such memories with the fact that Uncle Bruce's children grew into fine, hardworking adults who loved their own children. I ended up believing that the "love" in that proverb was more important than the "rod."

Dad never used a rod but often purposely selected and used a switch. One day, I did something that didn't give him time to deliberate. Whatever I did was enough for him to kick me and call me a little devil. Feeling that I deserved the kick, I soon forgot about it.

The next morning, Mom called me aside and said, "Dad wanted you to know he is sorry he called you a 'little devil.' You are a child of God."

I appreciated Dad's wanting me to know that, even though I already did. I had read it in the Bible and had been told so many times by my parents. Consequently, as Mom was talking to me, I was hearing his well-meant apology but pondering his absence. Why send Mom to apologize for what he wished he had not said? It takes empathy, honesty, and courage to thoughtfully apologize for something you wished you had not done.

Maybe that is why Coach Herrick's apology still sticks in my mind. I had sat on the bench throughout the baseball season except for the last game. Just before heading home, he came to the back of the bus and said he was sorry and that he had made a mistake not playing me more, a decision that kept me from heading off to college with a jacket silently declaring I had lettered in a high school sport (*EG*, 146–148).

> Mother and Bernie were over and cleaned up my house for me. Won't you come down and look over your future home and also get a peek at a lovely [bachelor] quarter. But I'm not so lonely. I am busy, with work in the field, S. S. work, visiting a few friends, not as many as I should (It seems I don't use my time right.) and then last and most fun and important of all; to write to the dearest little woman in the world. You don't mind if I call you a woman instead of a girl do you? Girl somehow makes me think of silly chatter and in the word woman, I see strength and steadfast, faithfulness and somehow; well something, that reminds me of mother and other mothers who are strong and comforting. You were worried about age, but don't even let that bother you. You had to be that age for me to realize the things I wanted in a wife. More than beauty or physical charm. Something age old and

eternal in its life. Love with hope and faith bound to it in such a way as to make of you as many of our mothers were, a comfort and a strength to their loved ones.

Typical of Dad. He respected women. I do not recall him ever disparaging Mom's comments or those of his daughters. Although my sisters usually helped Mom and my brothers and I usually helped Dad, comments such as "that's a job for a man" or "that's women's work" were not heard. My sisters often helped out with fieldwork, and my brothers and I often helped out with housework. My sisters learned to drive trucks and tractors. I learned to cook, sweep floors, iron clothes, mend socks, and use the Singer sewing machine.

Margaret I must go to supper. I also plan to visit a cousin, almost a favorite cousin and her husband and little boy. (5 months.) I love you too dearly to express it. But I pray for you and that we may soon be united in marriage and together in Christ.

In Christian love, Walter

May 22 [MAY 23, 4:30 P.M.]
My Dear Margaret:
"Pure religion and undefiled before our God and Father is this, to visit the fatherless and widows in their

affliction, and keep ones self unspotted from the world."
James 1;27

    This was the first Epistle text this morning and some how it came to me in a strong way. At least it caused me to look at it again and to ponder upon it. Perhaps because I fall short of doing these things and others, that I profess and confess to. I have always felt that it is necessary to be a doer as well as a confessor, and I hope that we together may work this out and plan our times to the best possible use and good. We will do this will we not?

    Are you well? Nearly ready to leave for Minnesota. I am glad for you and may your vacation be a happy and profitable one. I like to hear or rather read of your working upon things for our future home. I am so glad that you love a home and really and truly want one of your own. It will indeed be joyous to work in and share a home with you. That I know and look forward too. . . .

    I went home for a little while the other day, during a rain and mother said: <u>Won't it be nice when Margaret and Walter will come over when it rains or storms. I am sure we will get a cup of very good coffee, for mother can make good coffee</u>. They all look forward to meeting and seeing you, for except Bruce, they have not met you.

Monday Morn.
Dear Margaret,

"God is good and gracious, far beyond any measure that I know."

I know you would appreciate this statement more if you were here to see the wonderful weather we are having. I just heard, that already we have had within two inches of as much moisture as we had all last year. Normal is about 11 to 12 inches.

Never have I known a better spring, nor been up with my work as well as I am now. I have work to do, but I am not crowded as so often is the case. I do not think we will get water today because of the plentiful rains. They will put it in the reservoirs instead. I hauled a little straw for the horses and Mrs. Brown's chickens this morning. Then before starting anything else, I thot I would finish this letter in order to get it off in todays mail. It will probably be the last I mail to Washburn. There should be one waiting for you at Clarissa, Saturday. May your journey be pleasant and safe. . . .

You also asked about the distance between Minnesota and Colorado. It is called about 1000 miles from Denver to Minneapolis. I am fifty miles north of Denver and you are some west of Minneapolis, also some north though so I think it is about 950 miles from here to Clarissa. It is good road so I plan to drive thru when I come some time this summer. I will probably go by way of Omaha. . . .

This afternoon I hope I can harrow my beans. They will soon be up so I must harrow them soon if I am going to harrow them at all. This may be Greek to you, but I shall show and teach you all these things. . . .

About finances we will not worry. There is a man here who works for Browns and on the W.P.A., who supports a family with several children on $44 a month. I have plenty and am much better off than many. Our necessities will be provided for.

Now may our gracious Father in Heaven, guide, keep and protect you always.

In Christ, Walter

By the time my childhood memories were in place, Dad's father had passed away, and Grandma Johnson was living in Greeley with her daughter, Aunt Irene. We often stopped and left off some cream or eggs on our way home from church. On rare occasions, we stayed for lunch. During the Christmas celebrations at Grandma's, Mom tolerated Aunt Irene's dressing up as Santa's helper when the gifts were distributed, but on the way home, she would point out that there was no such thing as Santa Claus.

Dad always spoke lovingly of his parents. He said his dad was a strict disciplinarian, a hard worker, and well thought of in the community. I don't recall Mom commenting about my grandfather on Dad's side, other than to say that he got so drunk one time he drove into a ditch. No doubt he did, but I don't recall Dad ever mentioning it.

Mother-in-law and daughter-in-law relationships are often difficult. Paul, my older brother, remembers a day when Grandma Johnson and Aunt Irene dropped by our farm, and Grandma stayed in the car the whole time.

She passed away when I was in the first or second grade. I went with Dad to the funeral home. Grandma was lying on her back, eyes closed, facing the ceiling. Dad carefully adjusted her collar. I touched her hand. It was cold. There were no tears. A minute or so later, Dad said, "Let's go."

May 23 [MAY 24, 3:00 P.M.]
Dear Walter,

. . .

Thanks for the two air-mail letters. I do love to receive letters from you. I hope you will not be too busy to write to me this summer at least once a week. I certainly feel sorry about Mr. Brown's drinking of liquor. I am praying that he might see the sinfulness of it. I am so afraid of drunkards that I am afraid maybe I won't be of much help to you. God can give me the courage to brave it though. I can do all things in Him who strengthens me.

God didn't give Mom the courage, even though she had a good reason to brave it. Mr. Brown's farm was only a little over a mile from where Dad grew up. It would have been easy to move farm equipment back and forth over that short distance, even back in 1939. Yet Dad and she farmed the Brown land for only one year, or so I was told once I was old enough to be curious about the places our family had lived before arriving on the farm on which I grew up.

>I finally got my tan and green dress ready. It just fits me. Some of them said, "Walter, should see you now in your new outfit." God willing you may see it this summer. It is a dress that I can wear next fall also when we go to church together. The green hat matches the green in the dress.
>
>It surely would be fun to have you come to Washburn at noon on Friday. I am afraid though that your visit this summer would be more pleasant because right now I believe they are pretty busy at home without mother at home. I know, though, that the folks would welcome you. I am afraid you won't be able to stay very long and it is going to be awfully hard to say good-bye to you this time. Soon I am going to enroll in mother's sewing, cooking, and home managing course that she is offering to tutor me in this summer. I hope to be able to get far enough along in that course so that by the time you come for a visit I may prove my work to you. May God grant me strength for the wonderful future that He has in store for me.

Mom was a good student and later enrolled me in her many rewarding classes in sewing, cooking, ironing, exercising, singing, reading, and losing oneself in the endless fascinations of life and its understandings.

> You are soon to be busy watering. I shall be glad to carry the lantern for you. I hope that we will always live in Christian love toward one another. I don't fathom all this love which you shower on me but I drink in the knowledge that you do love me. I can't understand how you can because you don't know what I am really like but I hope you won't be too disappointed when you do learn to know me. Walter, I yearn for the time when we are to live together. I hope we can someday be knit together in love for one another and for others that we can win souls for Christ. I am tired now and I presume that this letter doesn't sound worthy of you so I just must stop writing and go to sleep. I hope to awaken early enough to complete this letter and then to get it mailed that it might go on the morning train.
> May God watch over us every step and minute of the way of our lives.
> Good night, Margaret

I expect Mom did hold the lantern for Dad a few times shortly after they were married, but that task fell on my brothers and me once we entered grade school. When I was awakened at night to help Dad reset the water by holding the flashlight, I knew my rhetorical question, "Do I have to?" would be answered affirmatively.

Riding with Dad up the field in the pickup always revived my spirit. I enjoyed seeing Dad pull the dam, seeing the water rush forward, hustling with him down the ditch, and then holding the flashlight while he quickly reset the dam before the onrushing water arrived. The night sounds, the ways of the water, and the hopes for the crop were engaging distractions. Whenever my flashlight wandered, I would be told, "Don't just wave the light around. Put it where I'm working." When using irrigation tubes, we would wait for the water to rise high enough to start them. Otherwise, we made small cuts in the ditch bank and shortly after adjusted them so the water was equally allocated to the rows. Once the water was set, we headed back.

Good morning!
"On the leafy treetops
where no fears intrude
God is ever good, ever good."

This little verse which Ruth and I used to sing in Sunday School is ringing true this morning. We have had a little sprinkle that no doubt cheers the birds.

Helen came home from the State Luther League convention very happy. She didn't return in time for the banquet and prom. I almost believe she had a vision or realization of the sinfulness of dancing so she just didn't appear on the scene. I hope so. Walter all the juniors and seniors danced except one senior boy, Rodney Slagg. *Off and on this winter they have held parties the sole purpose of which was to teach each one how to dance. It is disheartening.*

Yesterday I sent my letters for references and for the application of a life certificate. When I was out to Wilton, some ladies were talking to me, one told me "By all means get your life certificate. I made use of mine this year for the first time after marriage and it certainly has helped me out."

If God wishes to bless us so bountiful as to keep you by my side and to bless our work together then I won't ever need it but God wants us to work and I want to be prepared. Walter, I don't mean to be independent of you. I know that you will as long as God gives you strength [to] gain a livelihood for us two and[,] should God will[,] for our children. I am getting this certificate in case war should call our men to some foreign country then the women will have to be drafted in to take their places in the teaching profession.

. . .

Two of the teachers have had quite a time this winter. Different married men have asked to spend evenings with them. One of the teachers asked me last Sunday night if I could tell her the reason for the men having such a terrible opinion of her. God has spared me from experiencing such tragic experiences. God will ever take care of us and will lead us away from temptation if we come to Him and ask His help.

This coming Thursday I plan to send my trunk on to Bismarck. The bus that comes thru here to Bismarck is small so it could happen that they would not be able to take so much luggage on Friday. I presume other teachers will be going home on Friday. I am to leave on the 2:30 p.m. bus and about 10:30 the next morning. Soon I will be home.

Walter, I don't mind what you call me because I know that you will always call me names of respect and love.

I am enclosing two pictures. The girl with me is Blanche Holtan. The other picture is of the junior choir. In Christian love I want God to bless you always,

Margaret

I got my pencil last night. I have misplaced my fountain pen. I hope that I can find it sometime today. I need it very much.

Those same disheartening choices were made by us, her children. Mom didn't want us to go to the high school proms; Dad didn't seem to mind. When we discussed dancing, the best thing Mom could say about it was that Ebba, her oldest sister, loved waltzing when she was in college.

On entering college, I joined the marching and concert bands, much to Mom's satisfaction. But my musical interests often drifted to the jazz and big band sounds of her college years. When the opportunity arose, I joined a pep band and a second-string dance band. Mom never objected to my playing the dance music that was popular when she was in college but encouraged my practicing classical pieces by accompanying me on the piano.

Interests change with time. I put away my trumpet after college. Forty years later, after moving to Florida, I picked it up again, curious if my finding a physical exercise I could do for the rest of my life (*EG*, 63) applied to my embouchure as well. It did. Within a year, my embouchure returned. It didn't have the lip strength of my college days, but the feeling and sound were better.

The freedom of jazz still beckons, but now the great hymns of my church do as well. A few Easters back (before laying aside my trumpet because of a hernia), I was accompanying a choir anthem when Mom somehow floated in and proudly listened. At the end of the service, the congregation was invited to join in singing the "Hallelujah" chorus. I could sense Mom singing with the sopranos while I sang with the basses. It was a "heavenly" experience.

>May 25 [MAY 26, 4:30 P.M.]
>My dear Margaret:
>>Many, O Jehovah my god, are the wonderful works which thou hast done,
>>>And thy thoughts which are us-ward:
>>>They cannot be set in order unto thee;
>>>If I would declare and speak of them,
>>>They are more than can be numbered. Psalm 40;5

This letter finds you at home in Clarissa. I hope it finds you well and happy after completing your teaching. It would be nice if you could rest for a few days now. Can you do that? Are you going to direct the Choir this summer? It must be nice to be home again after being away for nearly five months.

You will have time to catch up with your music practice now won't you. I hope you plan to keep up with it. Even tho we may not be able to get a piano right away, we could and shall plan for one as soon as possible. I may not ever know much about music but I love to hear it and I will enjoy coming home evenings, to hear you sing and play, after we have finished our days work.

Please write and tell me of your homeward journey. Was the country beautiful and all green? Have you had much moisture in Minn.? Do the crops look good?

The crops look very good here and I understand Kansas has an excellent prospect for a bumper wheat crop....

I did not get the airmail letter until Mon. It probably got to Eaton Sat, but too late to go out on the route....

I am quite busy. Today I finished harrowing the potatoe ground and shoveled ditch this afternoon. Shoveling ditch is hard work, so I am tired. It was a waste ditch that I shoveled. We have to have a waste ditch to carry off the extra water that runs off of the land. I will also shovel all day tomorrow and I think I get the water tomorrow night....

I hope you and I can soon afford a radio. I do not like all the programs, but some are good and it is a pleasant change to listen to them. . . .

Now I say as you said: May we ever be ready for the Heavenly Bridegroom and may we ever walk as he would have us walk.

Yours in Christ, Walter

Dad enjoyed showing us how to shovel, and he was good at it. When shoveling weeds around the farmhouse, he would point out that if the shovel was sharp and you didn't go too shallow or too deep, it would slice through the roots with little effort. Dad thought of shoveling as an art and liked to say that he never had to lay down his shovel for anyone.

There is an efficiency, a grace, and an artistry to be found in almost any physical endeavor. When living in Kalamazoo, I came to experience shoveling snow as an art. The trick was to find a rhythm suited to the size and shape of the shovel and the depth of and moisture in the snow. When we had less than five or six inches, I enjoyed rhythmically scooping and throwing the snow from our driveway, first down one side and then up the other.

A Glimpse Back

May 17. Marital partners can lift each other up materially and mentally. Although both are important, it is not always clear just how it's done. In her May 11 letter, Margaret even wondered just what she did to inspire Walter. In this letter, Walter wants to help Margaret with her worries and promises to encourage her any way he can. How do you weigh the material and mental benefits of encouraging one another in a relationship?

May 20. In this letter, Walter tells Margaret that he wants them "to take any little differences that may arise" between them to God. How does the merit of their doing that depend on the nature and extent of the differences, their comfort in discussing differences, and their mutual understanding of God's involvement in their lives?

May 23. Margaret mentioned God protecting her from the experiences two other teachers had when married men asked "to spend evenings with them." What is it about our dress, demeanor, walk, and talk when meeting strangers that conveys opportunity in a relationship or lack thereof? Do we have auras, and do they have conscious aspects? If we do, where are they sourced, and how are they related to our spirits and goals in life?

A Helper Fit for Me

Walter can't remember a single time it was hard to speak to Margaret. We get a glimpse of how that came about for him in these letters. Here's how it came about for me.

During my second year at the University of North Carolina at Chapel Hill, I attended a graduate school picnic. I looked forward to plenty of food, a crowd of attendees, and meeting other graduate students—especially some I might date.

While in line for the food, I had started talking to a too-tall girl when I spotted my friend Dave Korts conversing with two other attractive coeds. At my suggestion, we joined them. Dave introduced us to Myrnice and Martha. Both were enrolled in a master's program in parasitology. As the afternoon wore on, attendees started heading for the evening dance. The girl I had first met decided to go back to the graduate dorm rather than to the dance. That didn't seem right, so the five of us went to a bar instead and had a great time.

Who to ask out the next weekend? Even though God had assured me I didn't know what I should be praying for when it came to a life mate (*QGW*, 137–139), I was still stuck on attributes. Myrnice had long dark hair and was from the North. Martha had short auburn hair and was from the South. I called Myrnice. We had a nice enough time, but nothing really clicked. So the next weekend, I called Martha—that would never have happened had I known she and Myrnice were roommates.

My asking a girl out for a date for the first time was usually a short and awkward affair. Lacking reasonable conversational skills in a tense situation, after the "Hi, how's it going?" I usually went right to the basic question: "Would you be interested in seeing a movie with me this Friday?" or something like that. And that's what I asked Martha.

"I'm sorry. I would love to, but I can't this Friday because my friend from college is coming this weekend" was the deflating reply.

Rejections were always hard to take. I had heard so many. I returned my stock reply, "That's OK, maybe some other time," knowing there probably wasn't going to be another time, at least not anytime soon.

"No, wait! I really want to go out, but I can't just leave my friend who came up for the weekend. Maybe you could see if one of your friends would like to double-date."

I could hardly believe it. *She was asking me out for a date!* Fortunately, I had a friend who was free that weekend. A couple of days later the four of us drove to Myrtle Beach. The whole day was a blast, ending with dinner at an oyster bar on Pawleys Island that Martha and her friend Minor had frequented.

Martha and I started spending quite a bit of time together. One day I was having coffee with Dave when I saw Martha walking by. I told him, "There's the girl I'm dating."

"You're dating her?"

I puzzled over Dave's unencouraging reaction until I learned why. Dave, who could beat me in chess while blindfolded, was quite particular in how he took care of things. Martha was a rather free spirit who laughingly said "oops" the time sugar went everywhere after simultaneously ripping open three packets for her coffee. On telling her about Dave's reaction, I learned how she and Dave had been assigned adjacent lab seats in a microbiology class and how irritated he had become when she accidently left an inky black fingerprint on his lab notebook.

Things didn't work out between Dave and Myrnice either, and they almost didn't work out for me a week or so later when Bill and Wendy, two other friends of mine, invited us over for dinner. It was a relaxing Friday evening that bubbled with pleasant conversation. They had a couple of extra tickets for that Saturday's football game and wondered if we cared to join them. We gladly accepted, but on taking Martha back to Ehringhaus, the graduate dorm, I began pondering the rightness of what I was doing. She had often expressed her love of the South. I was heading back west as soon as I received my degree, hopefully to Colorado. After dropping her off, I decided it was better to stop dating her before things got any more serious as a consequence of the hurt I caused by breaking up with a girl I had met a couple of months earlier who had quickly wanted to get much more serious than I did.

    The next day, Bill and Wendy picked me up, and we headed to Martha's dorm. When we arrived, I was still struggling with the best way to tell her after the game. I picked up the lobby phone and called her room.

    "Hello!"

    It was Martha's voice. "You ready?"

    "I'll be right down."

    A couple of minutes later, a radiant creature came bouncing through the door. She wore a short fall-plaid wool skirt. Her auburn hair glowed. She was bubbling over with expectancy. In that moment, I realized *I'd be a fool to walk away from my joy.*

    That afternoon, I got lost in Martha and have pretty much remained so ever since. We could talk about whatever came to mind. Well . . . almost. A month or so later, we were having a late Sunday-morning breakfast in an intimate booth at our favorite coffee shop, the Carolina Café. Classical music played in the background. I was excitedly sharing the essence of a couple of insights regarding a recent theorem on stochastic approximation for my dissertation when, out of the blue, she said, "If I hear one more word about statistics, I think I'll scream."

Didn't matter. Talking about statistics was not what brought us together, and there was everything else in the world to share.

Ours was a coming together in which we both came to see the workings of God. Of course, the physical attraction was there, but largely through her reticence, it played little role before we decided to marry three months later. That was well over fifty years ago. We're happily married and best friends—soulmates with a deep relationship that has negotiated many storms while sailing the rich and beautiful waters that have come to define our life together.

Later, I spoke at the wedding ceremony of my daughter, Rebecca. I wanted to tell her and Matt, her soon-to-be husband, what being a marital partner meant to me. Genesis 2:18 came to mind: "The Lord God said, 'It is not good for the man to be alone; I will make him a helper fit for him.'"

I was curious how that word *helper* was used in the Bible, so I got out *The NIV Exhaustive Concordance*. It contains the 9,597 Hebrew words used in the Tanakh, more widely known by Christians as the Old Testament. The Hebrew word translated as *helper* in the above verse is number 6,469 in the *Concordance*. It occurs more than twenty times in the Tanakh, never referring to a hired hand or a human assistant. In Psalms 89:19, it refers to the strength God has given a warrior. Otherwise, it refers to God or God's help. When thinking about the "helper" God made for me, my mind often jumps back to the lesson I learned when God answered my prayer by bringing into my life the "helper" who rescued me from my occasional and almost suicidal bouts of loneliness (*QGW*, 186–187).

CHAPTER 3
DR. PIHLBLAD

The following are some of the headlines splashed across the *New York Times* as Walter and Margaret were deepening their correspondence. June 2: "Chamberlain [Britain's prime minister] seeks mediation in Spain." June 9: "Canton is ablaze on new raids, casualties 8,000." June 10: "400 planes bought by Britain in U. S." June 12: "U. S. acts to check sale to Japanese of our war planes."

In Lindsborg, Kansas, where Margaret went to Bethany College, readers of the *McPherson Daily Republican* learned on May 24 that Dr. Pihlblad, the head of the music department, was critically hurt in a car accident. In the *Lutheran Companion*, Walter and Margaret learned that a synod commission had addressed "Gambling, Labor Relations, and Liquor" and that the Annual Convention of the Augustana Synod would be held in Brooklyn, New York, June 20–26.

In an upcoming letter, Margaret imagines a trip she would like to take with Walter that would give rise to a trip I was able to take with them to see the people and places discussed in their letters.

The May 27–June 15 Letters

The Metropole Hotel, Fargo, N. Dak., May 27, 1938
>My dear Walter,
>"Jehovah will keep thy going out and thy coming in
>From this time forth and for evermore." Psalm 121:8

For this God is our God for ever and ever:

He will be our guide even unto death Ps. 48:14.

Now my school days are ended, God willing. It hardly seems true. I am so glad that I have been able to teach now for five consecutive years. I have such a bright future—to think that I will have you near me to help and guide me and to supply my necessities[.] God grants you his grace.

I can hardly wait to get to Clarissa—then I will have a letter waiting for me from you. I wish you would be there in person instead. When do you plan to come to Clarissa? Oh, it will be enjoyable to have you with me.

We had a wonderful graduation commencement exercises. The auditorium was just packed. The mixed chorus sang very well. It made me feel a little sad to think that that would be the last time that I would hear them sing. I pray that each member may sing in the heavenly choir. . . .

Soon I'll be home! Won't the folks be eager to see the (our) diamond ring and your photograph? Walter, I surely thank you for them. I am still pondering about what we should buy for the ten dollar bill you sent. This morning when I was packing I thought a suitcase would be just the thing. I have been using my father's suitcase except for a small one which I received for graduation. When we leave Clarissa next fall, it will come in handy. Then, too, since both you and I love to travel, we can someday make use of it again. . . .

Walter, I think it will be fun to come to your parental home. May God grant us to have such a home as yours and mine that when, if God permits us to have children, they are grown as we are they will long to come home too. . . .

I left the $130, that I had saved, in the bank. I will not be tempted to draw it out this summer if it is [in] Washburn. I paid all my bills in Washburn and I have about seventy-five dollars left to last me this summer. My room rent for tonight is $1.25. I have a nice room. My fare comes to $7.10. I really have enough clothes to last me this summer. . . .

Say, if you get sheep instead of cattle we will get to have roast lamb sometime. I am afraid of cattle but not of sheep. It is certainly working out fine. Mother's folks raised sheep.

Now may God richly bless you with many spiritual and temporal blessings. I hope that it won't be long before you come to Minnesota. I am sure that your car will enjoy taking you up to the Northland.

In Christian love, Margaret

"All things come home at eventide,

like birds that weary of their roaming

And I would hasten to your side, homing."

May 29 [MAY 29, 4:30 P.M.]

My Dear Margaret:

If God is for us, who is against us? Rom. 8:31. . . .

This verse which I am sending to you, has come to my mind many times but I could not think where it was and remembered to ask anyone who might know. Last night in preparing my S. S. outline for today and also my lesson I came across it.

The letter to the Rom. is quite deep is it not and harder to understand than any of the others. It does not seem too difficult tho and it is not dry or uninteresting by any means. I hope that we can do quite a lot of studying together. I think if two people can study together that it goes faster and easier [and] also becomes more clear because one sees where perhaps the other does not. Please keep up your English and grammar. I believe them to be a big help in study, in order to get the meaning more clearly. Have you had any language? Am I mistaken or did you not say that you taught Latin one year?

Well, I have the water now and do not get much sleep, 5–6 hours on the average. I like to irrigate though. It seems you can almost see the hay grow, as you give it water.

How I do wish my Margaret could have seen the sunset I saw last nite. I cannot describe it but it was beautiful even tho it was wild. There were just a few smoke clouds in the sky being driven by a strong wind in the upper atmosphere and they took on some beautiful colors, some very red and altogether it was the wildest sunset I have ever seen. . . .

How I wish you were here to cook breakfast for me and eat it with me and boy am I hungry. But that's enough wishing. I have been up two hours and a little over now at 5:45 so I am going to get an hour's sleep. Then up, shave, set the water, breakfast and off to S. S. That's a kind of picture of farm life, but I like it. Especially if we are careful I know we can arrange our time for some study and reading, especially in the winter. . . .

Yours in Christ, Walter.

The question in Walter's chosen verse flows out of a life view in which a loving God is ultimately in control. At the same time, who would question the criticality of the "If"? To take a thought from Shakespeare, therein lies the rub. Is Paul referring to an outer spirit that spawns galaxies yet permeates the smallest of atomic particles? Is he referring to an inner spirit whose personal direction can be felt? Might he be referring to the endless coincidences by which things come about and an inner joy in seeing them? He is talking about something profoundly affecting his being, but can we ever agree on exactly what that something is?

Clarissa, Minnesota, May 31 [MAY 31, 3:00 P.M.]
Dear Walter,
"And in none other is there salvation: for neither is there any other name under heaven that is given among men, wherein we must be saved." [Acts 4:12]

Yesterday morning in church John spoke on "Fear not therefore: ye are of more value than many sparrows." It gave me strength in faith. I had the privilege to sing "For His eye is on the sparrow and I know He watches me." It went so easy to sing. I don't believe I have ever sung as well before. I am so happy that God has given me such a true Christian who is some day going to be my husband so maybe that is the reason that it went so well. . . .

Your letter came the same morning as I came. I surely enjoyed receiving the letter. I surely would enjoy being in Greeley to attend the Luther League convention, but I don't believe I can from the financial stand point. The money that I had John needed so badly to pay some debt of our family. Bernard had no money to send but he said he would have some in July the latter part so then I will receive my money back again. . . .

You asked me about my trip home. It was enjoyable to travel because the fields were very green, the lakes brimming over with bluish colored water, the roads were smooth, and the people were congenial. . . .

I remember Arthur Francen and Jeanette Elder. I hope that the man you mentioned who is to visit in Sweden stays quite a long time so that they can be settled in his home for a longer time that they are planning to. When Arthur and Jeanette went to communion which one went up first? . . .

A silly question regarding marital politeness? Maybe so, but Margaret ended up with a man who let her lead the way to and from the communion table, loved and respected her, and never left her trailing behind.

Now may God bless you and keep you always. I am waiting for you.
Yours in Christ, Margaret

In Margaret's selected verse, Peter puts in his own words something Jesus shared with his disciples during their last supper together on the night Jesus was arrested. From John 14:6: "I am the way, and the truth, and the life; no one comes to the Father, but by me." The promise of a well-chosen way begins with the first step; the joy that comes with that choice often does as well.

But who was Jesus? He was at least a spiritual pathfinder whose life changed, and is still changing, the lives of many. He was willing to give his life for a kingdom without swords before many of his followers would be willing to give theirs. He died forgiving even those who took his life so they too might experience the joy and peace of the way to God that he taught, lived, and still leads.

So how did Jesus see "the way"? Although it is understandable that the followers of Jesus would have different understandings of something so rich and multifaceted—and often troubling—as the way of or to God, I doubt his followers ever felt he was intentionally ambiguous about "the way." They quote him in John 14:21 and 23 as saying, "He who has my commandments and keeps them, he it is who loves me . . . If a man loves me, he will keep my word." Thus, the question becomes, What were his words and commandments? All we have of them are what his disciples and followers lived, remembered, and wrote long after he died.

May 31 [JUNE 1, 4:30 P.M.]

My Dearest Margaret:

> All the way my Saviour leads me;
> What have I to ask beside?
> Can I doubt His tender mercy,
> Who through life has been my guide?

I cannot thank God enough for His leadership. Especially when I think of the many things you and I have in common. . . .

Frank Lunn came down from Cheyenne Memorial Day and will stay until tonight at least and perhaps tomorrow night. . . . He went down to Greeley to visit Rueben and Evodia a little while. They are both interested in the affairs at Bethany. . . .

Have you heard that Pihlblads were in an accident and that they were injured very seriously. He was injured Monday and was still unconscious Fri. when Frank left. . . .

I do not know now when I will be able to come to Minn. not for several weeks yet. These coming 6 weeks are our busiest ones. Potatoe planting [and] haying. Potatoe and bean cultivating, cutting and thrashing barley and weed cutting.

I hope you will be busy so that the time will fly swiftly. If you have time would you like to send me a kind of outline for the wedding or plan if we should call it that. . . .

I'm waiting for Frank to come for me, to take me home to supper. He took my car to Greeley so I am afoot. He should be here anytime now; so I will close this short letter. . . .

In Christian love, Walter.

P.S. You have managed well to save so much money.

May 31 [JUNE 1, 3:00 P.M.]

Dear Walter,

"Count you many blessings

Name them one by one." . . .

Tomorrow is washday. Ruth is to come to help us. It won't take very long but it will be a relief to have it done. We have had such wet weather that it has been impossible to get clothes dry. Since the sky cleared up before the sun set, I believe we will have ideal washing weather.

You certainly aren't getting much sleep. Can't you sleep some during the day? You could set your alarm clock so that your irrigating wouldn't need to suffer. You need more rest. . . .

Ebba just told me that the theatre in town has been changed into a miniature old peoples' home. You'll have to see how neat the home looks. Different people have tried off and on to run a theatre here but they have always been unsuccessful. Mrs. Samuel Johnson told Ebba that some years ago <u>there was a group who held street meetings in this town who prayed that there would never be movies shown here. Prayer prevails</u>. . . .

In the last Lutheran Companion there is an account of the terrible accident that happened to President Pihlblad, his wife, and Jens Stensaas. I wonder how they are getting along now. Conditions in Bethany are quite sick. Adelyn told me somewhat of why Uhi was told to resign. We need to pray about all this.

You asked me if I have had any language study. I have but I haven't made much use of it. In high school I studied Latin and French; in college I studied Latin and German. At home we have all learned Swedish. I want to continue studying Literature and grammar after we get settled in our cute little home. . . .

In Christ, Margaret.

It is easy to be facetious and say that God answered the prayers of the Clarissa citizens by changing a theater for the young into a "miniature old peoples' home." But I'm more interested in the power of prayer in bringing about a more rewarding view of life. What follows is a personal experience.

The need arose when Martha and I found ourselves on different sides regarding the pros and cons of taking our two children out of public school and sending them to a private Christian school. We both saw educational, financial, social, and spiritual issues at stake.

Each time we sat down to eat and discuss our children's day, I could feel my stomach tensing around an inner resistance, regardless of how carefully Martha broached the issue. Through the months, we descended into an argumentative quagmire that started the proverb "better is a dry morsel with quiet than a house full of feasting in strife" drumming in my mind. My "yes, buts" slowly congealed into a seemingly impenetrable wall for Martha and a discomfiting inner lead ball for me. That weight was not to be wished or rationalized away by the "rightness of my principles." It kept growing until I stumbled upon a simple prayer that I would silently utter whenever I felt my stomach begin to tighten with another "yes, but": "Dear God, please take this lead ball from me so I can hear straight."

As the "yes, buts" tapered off, the lead ball crumbled and eventually dissolved. New thoughts and exchanges along a much more basic theme arose—not who was right or wrong, but who was best positioned to decide. While I was at work enjoying my professional day, Martha was getting the children ready for school and addressing their concerns when they returned. She was helping in the classrooms. She was discovering the destructive behaviors that troubled us, and the resolution of which went beyond what the dedicated but overloaded school staff had the will and the means to address. The argument was amicably resolved within a couple of weeks.

Based on the outcome, one might argue that God was on Martha's side. That was not how I looked at it. I saw that God had led me to a readily answered prayer, a plea for help in silencing my endless stream of "yes, buts" so that I could hear whatever I needed to hear. Why would a loving and merciful God not answer such a prayer?

I also came to a better understanding of the commandment, "You shall not take the name of the Lord your God in vain; for the Lord will not hold him guiltless" (Exod. 20:7). The passage puzzled me in my youth. Why would the Creator of the galaxies hold me guilty if I, nothing but a child, misused his name? In spite of my question, I didn't. Reading the books of Moses at my young age profoundly impressed upon me the holiness of that name. Thankfully, when my need came, that name was there and extricated me from my entangled arguments.

June 3 [JUNE 3, 4:30 P.M.]
    My Dear Margaret:
    His work is honor and majesty;
    And his righteousness endureth forever.
    He hath made his wonderful works to be remembered:

Jehovah is gracious and merciful. Psalm 111:3,4

I love Jehovah, because he heareth my voice and supplication.

Because he hath inclined his ear unto me,

Therefore will I call upon him as long as I live. Psalm 116:1,2.

Margaret, I do not see how anyone could say there is no God at all, after beholding this earth and all its beauty and most especially this morning. It is a privilege and we, who are not only alive, but have also all our faculties, such as sight, hearing, thot and motion, must certainly thank God without end. . . .

It was nice that you could help your brothers and you need not worry either about coming out here, buying any furniture or even an elaborate wedding. I say truthfully that I like simplicity. Only, that we be in God's temple and married by one of his workers. This will certainly not be hard for us. I am glad that you have brothers who are ministers. It is a wonderful calling.

I told my folks a long time ago that we should like them to come to the wedding, so they will have an opportunity to plan for it. . . .

I took Frank Lunn to Cheyenne Wed. afternoon and the plains between here and there are beautiful. They are very green with millions of flowers upon them. The Colorado primrose is quite a large flower (they call it the sand lily) and there were acres and acres of them.

I get the water again tomorrow. (Sat.) I only got two days rest, but I need the water so I do not care. . . .

I am happy in the fact that I am so busy, yet nevertheless it is hard to stop writing for it seems then that I must part from you again. May the day hasten, when that will not often, be necessary.

In Christian love, Walter

June 4 [JUNE 4, 3:00 P.M.]

Dear Walter,

"Everyone therefore who shall confess me before men, him will I also confess before my Father who is in heaven." Matt. 10:32.

This is Saturday morning. I awakened before the alarm went off so I am up before Ebba is. The sun is shining brightly and the birds are chirping. There are quite a few things planned for us to do so I thought I wanted to write to you before we have plunged into that work. . . .

Day before yesterday Lowell Anderson, a music pupil of Ebba's, mowed our lawn. Ruth, Ebba, and I tried to dig all the dandelions before he mowed. It was quite a task but lots of fun. The mosquitoes fought against us to help the lions, I presume. They seemed to attack me more severely than the others. I was bit on my face, neck, arms, and hands. You should have seen my swollen face. I put glycerin on it and now the marks are disappearing. Yesterday I picked 210 dandelions. The mosquitoes were kinder to me then. They must have had a consultation concerning their mistreating me.

The funeral of one of my former pupils is to be held this afternoon in Eagle Bend. This boy, Clarence Peterson, was such a gifted boy in scholastics and was a well-behaved pupil. I can hardly realize that he is not living any longer. About a month ago he became ill while working in the field. He was given medical aid immediately and the doctor diagnosed it to be spinal-meningitis. He had been so healthy and just all of a sudden he became ill. "We know not what awaits us. God kindly veils our eyes but every step of our onward way He makes new scenes arise and every joy He brings us comes a sweet and great surprise." I hope he was saved. I am to take the star route to Eagle bend at noon because I am to sing the song "He Knows." I quoted a few words of the song in this paragraph.

In less than eight weeks I can be expecting you to drive up on our lawn. <u>I wish that I could meet you in Minneapolis and from there we could go to Mora and on up to Duluth and back here to Clarissa</u>. That might be too long [and] round about [of a] way. I haven't looked it up on the map but the thought just came to me. . . .

I am sure you have enjoyed having Frank Lunn with you. I hope that you got the chance to take him to Cheyenne. I know that Evodia and Reuben appreciated hearing about Bethany's affairs. I received a letter from Miss Hanson and she wrote that everything seems dark with [the] Pihlblads and Stensaas so tragically hurt. . . .

Walter, whenever I think of our wedding I get thrilled clear thru but then I get a little afraid too. I hope that mother will sew my dress. I want the dress to be white in color. I wonder what time, in the morning after services, in the late afternoon, or in the evening the ceremony should be held. I rather like the second one. I don't think we should have a dinner after the wedding; a lunch would be better. There are so many to choose from to perform the ceremony, John, Reuben, or Philip. I would like to have mother give me to you. Since it will be hard for Bernard to break away, I don't believe he would be able to come. I haven't decided who is to be maid of honor either Ebba or Ruth. I would like to have Irene [Walter's oldest sister] be one of the bridesmaids. I don't believe we will have more than one bridesmaid. Adin would make a pretty good ring-bearer. The three little girls, Rhoda, Katherine, and Anna, would make sweet little flower girls. I wonder who you want for best-man and usher. It will be fun to discuss the wedding when you are here. May God be our helper in making our plans.

Yesterday Irene and I baked cookies, the day before Ruth and I baked buns. Ruby said for me to come down there and she would give me lessons in cooking. She wants me to learn to make good gravy because she said, "All farmers like good gravy." . . .

Now I wish you speed to come here. May God prosper your endeavors on the farm and in your church work. May His peace be yours.

In Christ, Margaret

P.S. The car that you and I rode to Julotta is seen quite often by our house. . . . I get a terrible longing to see you when I see the car. I am glad that you know how to drive and care for various cars. You mentioned one time that you would teach me to drive your car. I hope that I can learn well.

June 5 [JUNE 5, 4:30 P.M.]

Dear Margaret:

For he will give his angels charge over thee

To keep thee in all thy ways. Psalm 91:10

Good morning Margaret, how are you, this morning? Here in Colo. the sun is shining brightly with just a few clouds in the sky. . . .

Frank brot down a little radio and set it up for me. It is a Philco and it costs 15 dollars, wholesale. I plan to buy it. It is a good radio he says. He is a radio expert, in fact makes his way thru school repairing radios. Later on if we want a larger radio we can trade this one in, can we not? . . .

I certainly would like to see you. . . . But now I must close; happy that you are in charge of one who wants to keep you for himself and me.

In Christian love to My Margaret. Walter

June 6 [JUNE 8, 6:30 A.M.]

My dear Walter,

"For the Son of man also came not to be ministered unto, but to minister, and to give His life a ransom for many." Mark 10:45.

This is Monday morning. I am sitting in the kitchen near the range. Ebba has gone to teach Bible school and, since Ruth has not come from Eagle Bend yet, I am all alone. I thought it would be fun to visit with you.

I have cleaned the kitchen pantry, back-porch, and made the bed so if someone comes to spend a while with me I don't have to excuse the way the house looks. . . .

Walter, I had such a wonderful Sunday yesterday. Quite a number of times though I wished so that you were with me. This waiting isn't as easy as I thought it would be. . . .

While at Gustafson's, John learned for a certainty that he is to go to the synodical convention in New York. . . . I would like to go because I do enjoy being with such a large Lutheran body and then I have never been to the East. . . .

"All the Way my Savior Leads Me.

What have I to ask beside?"

John and Ruth just came and now have gone. Ruth has gone to buy groceries and she said there was a letter from you to me. I can hardly wait to read it. Then John said that I could go along to Mora tomorrow and to the synod next week. Walter, I am so happy. May God use me in His own way. I'll let you know how the trip goes all along the way. God will direct us and will keep us safely on the way so that no serious accident occurs if it is His will. Maybe something will happen so that I can't go but I hope not. "Where He Leads me I Will Follow." I love to ride with John because we have prayer sessions as we go. . . .

Mother is feeling fine. She said she would sew my wedding dress if we girls would help her. She has been out calling on a number of families. She wants me to spend some time here, but she doesn't know about our plans of going East. John just told me that I was to go along as far as he knows. It is all if God wills it.

Phillip and John are going to the post office so I want this letter sent.

May God bless & keep you for me. I want to see you as soon as He wills it.

In Christ, Margaret.

June 9 [JUNE 9, 3:30 P.M.]
Dear Walter,

"Now the God of hope fill you with all joy and peace in believing, that ye may abound in hope, in the power of the Holy Spirit." Romans 15:13. . . .

Since there are certain reasons why we should not go to the synod I understand John is going to give up his chance. This isn't definite, his declining, but the way it looks there will be no trip. It is all right, but I surely would enjoy going. I have travelled a great deal but it seems I never tire of it. It takes money to travel and when one's funds are low and really should be used to pay debts then one must forego interesting trips.

I surely am glad that you are intending to buy Frank's Philco radio. John said that that kind of radio, Philco, is a good make. Won't we enjoy listening to good radio programs? I don't like to have the radio on from morning to night. I do like to hear good music and I like to have good lectures, too. Won't it be fun to choose the better programs from the list? . . .

We received a long letter from Bernard yesterday. He said that Dr. Pihlblad is no longer unconscious but his memory is hazy. Next Monday Mrs. Pihlblad is to have her leg set. We need to remember them in prayer. Think how painful the bruises must be. It seems that God had taken the affairs of the college in His hand and He doesn't want Pihlblad to have anything to do with them for a while. . . .

In Christ, Margaret

Walter led off his December 8 letter (*EG*, 88) with a concern about God's mysterious will in Cora Brinkman's death shortly after birthing a child. Margaret, who had a sister die in birthing her baby, did not care to comment. However, after hearing that the Pihlblads are getting better, Margaret sees God's hand in resolving an affair at Bethany. She first referred to it in her March 14 letter when she learned from her brother Bernard about a "non-Christian" affair involving Professor Uhe, the violin instructor and orchestra director at Bethany, that was causing quite a stir in Lindsborg.

>June 11 [JUNE 11, 12:30 P.M.]
>My Dear Margaret:
>
>. . .
>
>I hope you get a chance to go to the synodical meeting or Conference. I should like to go. Conrad Floreen will be ordained. He was student pastor at Cheyenne a few years ago and a fine man and friend. It will be a fine trip, besides as you said; a pleasant and wonderful experience to be together with such a large Lutheran body.
>
>I must close this short letter now. It is 5:30 a.m. and I must harness the horses and go to breakfast. I planned to write last nite, but Leonard Molander came over and talked for a while. I was too sleepy when he had gone so I arose this morning in time to write. I miss your letters so much even if the space is only 3 or 4 days. Now it is almost a week since I

wrote last. Please forgive me Margaret. I love you dearly and long to see you and be with you. . . .

Sunday afternoon I will stay at home to rest and then I will write for there is much more to write.

In his keeping, I leave you,

In Christ Walter

June 12 [illegible postmark]

My dear Walter,

"I am the true vine, and my Father is the husbandman. Every branch in me that beareth not fruit, he taketh it away : and ever branch that beareth fruit, he cleanseth it, that it may bear more fruit." St. John 15:1.

These words were the text that Philip used in his sermon. *He made it very clear to me how Jesus' body had to be cut to graft us into Him.* It is wonderful that God works to cleanse those, who are grafted in, that they may bear more fruit.

I am sure you have been very busy or else you haven't known my whereabouts or else you are sick. I haven't heard from you for a long time, or maybe I am just impatient, so the time seems long. I know it isn't easy to write if you're tired and weary, but I do think it fun to receive letters, so I am going to.

Saturday Edna and I baked buns and cinnamon rolls. Do you like cinnamon rolls? If you do I will practice on baking them. . . .

John wrote a card to Philip stating that according to plans now we are to go to the synod on Thursday morning. Philip is to take me to the cities on that morning and then I am to ride on to New York with John. It seems so wonderful! God will frustrate John's plans if He doesn't want us to go. . . .

Good morning, Walter. This is such a beautiful summer morning. I have felt like taking a walk out in the fresh air but I guess I will wait until this afternoon so that I can take Adin and Gerald with me. . . .

In about four weeks I can be expecting you. I hope that you will get a chance to rest up before going on such a long journey. It is difficult to drive when one is tired. Bernard said when he drove up to Minnesota that whenever he felt sleepy he just drove to the side of the road and slept for a while and then continued after he had gotten his rest. . . .

Which highway is our home by? I would like to see it on the road map. I believe you said our place was about 2½ miles from Eaton. In which direction from Eaton? . . .

Yours in Him, Margaret

The farmland around Eaton is largely laid out in a grid with roads a mile apart. The farmhouse in which I grew up was three miles east and two miles north of Eaton.

June 14 [JUNE 14, 4:30 P.M.]
My Dear Margaret:

. . .

I said in my last letter I would write again soon probably Sunday. Sunday morning Rueben said he was going to Haxtun and had no one to go with him except the two Sand girls. His car is old and he [felt] that if something went wrong with it, it would be better [to] have another man along. I had planned to stay at home and rest, but I decided to go along. For that reason I am late with another letter. I don't know what is wrong with me. I must not use my time right or I try to do too much myself.

At any rate I wanted a letter off this morning so I got up in time to write before breakfast. I put the horses in first. It is so fresh and fragrant this morning, with a heavy dew upon the ground. . . .

I am glad that you can help your brother, Philip and Mrs. Leaf as well as Gerald and Adin. I hope you are busy and from your letter I gather that you are. <u>It makes the time fly</u> and it is also the proper way for us to live, for there is plenty of work to be done in God's kingdom. Don't you think our tasks would be lighter, if we did every one, as tho we were doing it for Him.

Again I must go, but this morning I asked God to show me the proper use of my time and as he wills it, I shall write. I hope to write more often for I know that even a letter means so much. Forgive me my neglect and pray for me. Perhaps it is not right for me to spend so much of my time working at the things of this world, I mean so many hours at farming. If I had more faith, perhaps it would all go just as well. What do you think. Of course it is this way only for a few weeks at this time of the year.

Now I do not know whether I made myself clear in that paragraph or not, but one thing is clear. I love you and shall try to write at least twice a week. I know I certainly miss a letter from you if the interval is too long. . . .

In Christian Love, Walter

June 15 [JUNE 16, 4:30 P.M.]
My Dear Margaret:

I know that thou canst do all things,

And that no purpose of thine can be restrained. Job 42:2

I too, know that all things are and can be done, by our gracious Father in Heaven. I know that during the days past and the days we are now in, he has given me far more than I had dared hope for except as I hoped in and thru him.

Only God made it possible for me to meet you and for you to love me as you do. And only he would cause all things to go as well as they have been and are doing. I refer to my work and farming. For everything is working out just right it seems. True; I am very busy and I must say, working pretty steady if not hard. But that is proper anyway and I thank God, for health and strength to work. Never have I worked so hard and never has it been so easy to work. Knowing that I am working for you, and knowing that your prayers go to God, on my behalf; certainly makes it easy to work. . . .

Boy do I like cinnamon rolls and any fresh bread. I liked your mother's bread. Ill slip in often when you bake and get some fresh bread. But right now none of these things would matter if I could see you and put the ring upon your finger and kiss you again. Then won't you sit in my lap while we eat this roll together?

I am glad that you prayed as you worked. I often do, even tho it is only in the mind, God hears and understands, and I think these are often our most sincere and most heard prayers, because they are so spontaneous.

If you go to New York, I will be praying that He will guide you and keep you, going and coming. Of course this is always my prayer. I don't know when this letter will get to you if you go. If it stays in Minn. it will keep until you get back and if you stay for several days, it will be forwarded to you.

Our place is about 3½ [miles] from Eaton—East. . . .

Now I will mention a thot that has been running in my head for several days. If I sent you a ticket would you like to come out here and visit. You could stay both at my home and at Reuben's and Evodia's. You could meet my folks, and we could look over the house and make plans. It will need some cleaning and work done on it. Altho I doubt that Mr. Brown would spend a great deal, I think he would do quite a bit for me. Then you could chose the colors and make suggestions. We could even plan and look for furniture, perhaps even pick it out. This is merely a suggestion and God will direct us. It would cost far more than a ticket to drive to Minn. and at least that much if I went by bus.

Now the day is over and I must sleep some. Goodnight.

Now may He keep you, forever and ever.

In Christ, Walter.

A Glimpse Back

May 31. In commenting on the changing of a movie theater into a miniature old people's home, Margaret noted that prayer prevails. That letter was followed by an illustration of how a simple prayer led me to see my responsibility in a deep difficulty that had arisen in Martha's and my life. In what creative spirit do you seek guidance when difficulties arise in your relationships, and what underlies your confidence in that guidance?

June 3. In contemplating "this earth and all its beauty" and the faculties whereby he does, Walter wonders how "anyone could say there is no God at all." Do you feel there is a creative source that underlies that beauty and the faculties whereby it is perceived? If so, what words would you use to describe your creative source or to call upon it to guide your life?

June 4. What do you feel went through Margaret's mind in choosing the song she sang at her former student's funeral? How do you see the presence or absence of that student in her mind?

June 6. Regarding her going or not going to the synod meeting, Margaret says, "Where He Leads Me I Will Follow." To what or whom do you turn when wanting to make a choice that is "right" for you?

June 12. Margaret mentions that by virtue of Jesus's body being cut, she has been grafted into Jesus and cleansed by God. "Being grafted into the body of Jesus" and "being cleansed by God" are religious metaphors that bear on who Margaret is and how she is being made whole. What phrases would you use to describe who you are and how you are being made whole?

June 14. What exactly is the "it" in Walter's statement "It makes the time fly?" and in what way does it make time fly?

June 15. Walter is glad that Margaret often prays while at work. He feels those unspoken prayers are important and heard by the creative spirit because of their spontaneous and sincere nature. What experiences would lead you to agree or disagree with him.

A Trip with Walter and Margaret

On reading "I wish that I could meet you in Minneapolis and from there we could go to Mora and on up to Duluth" in Margaret's June 4 letter, I decided I would have loved to have joined them. Of course, I couldn't; I was yet to be born—or so I would have once mistakenly thought. Following the Night of the Little Self, I became increasingly aware of how much of my parents' spirits and interests reside in me and I in them, even after their physical deaths. With that awareness, I left Michigan a few summers ago to spend a couple of weeks with *Mom* and *Dad* in order to browse some of the newspaper headlines and explore some of the places mentioned in these letters.

I would be crossing the Great Plains but without Martha. She chose to head south to enjoy a high school reunion and visit her greater family in North Carolina rather than travel a "treeless expanse" for two weeks. With my Honda Helix strapped to the back of Buggles, our Safari Trek motor home, I made for Jefferson City, in the middle of Missouri, to have lunch with David Carlson, Uncle Reuben and Aunt Evodia's son, and his wife, Lois.

Once in Kansas and headed to Lindsborg, Mom's reminiscences about Bethany began commandeering my thoughts. I could feel *her* excitement in once again walking across the campus that had been so much a part of her life. *We* parked Buggles at a convenient RV park a few miles out of town. The next morning, I hopped onto the front of my Helix, *Mom* hopped on back, and *we* took off for a day together in Lindsborg.

We were looking forward to walking the halls of the music building where *Mom* had mindfully developed her talents for singing and choir directing. I could understand *Mom's* enjoyment in the weekend festival, which disappointed me. I had hoped to find some news clippings concerning the scandal in the music department that Dr. Pihlblad's car accident exposed, but the library was locked up. *We* drove around a little while before heading south to nearby McPherson. Fortunately, its library was open and had the desired headlines in some 1938 back issues of the *Daily Republican*. After scanning them, *we* returned to Lindsborg, where *we* met Gerald, Uncle Philip and Aunt Edna's son, who had spent the summer working on our farm after his high school graduation.

Gerald took *us* on a pleasant road trip around Lindsborg. After seeing the last parish where Morfar, Mom's father, had preached, *we* walked around the Leaf memorials in the Lindsborg cemetery. I enjoyed recognizing some of the names and dates, but they had little to do with Mom's notion of heaven and were largely irrelevant to my sense of *her* life after death. In these letters, she shares the "rejoicing in heaven" she expected once "Jesus took her home." The gravestones and dates reminded us of some with her in heaven, but we discussed what of their lives now lived in ours, not what of their lives now lay beneath our feet. After saying goodbye to Gerald, *Mom* and I stopped in to see Ruth Ann, Uncle Bernard and Aunt Judith's daughter. Then *we* headed back to Buggles to prepare for an early start on *our* trip to Colorado.

The rising sun cast the shadow of Buggles and its two big side-view mirrors on the flat road ahead as we headed west. *Dad* made his presence known when *his* ever-mindful spirit spotted a grain elevator on the horizon. My mind raced back to when Dad spotted one like it as he drove our late-forties Desoto Suburban across Western Kansas on our way to visit Mom's brothers and sisters. He rhetorically bet it was at least ten miles away even though clearly visible. I took *him* up on that bet a second time. I was a bit disappointed that *he* was again a mile shy, but I enjoyed the bet. After passing a few more of these elevators, *we* began the long climb to the Mile High City, where I visited two of my brothers and a sister and their families.

Greeley was *our* next stop. I scanned a few relevant news items from the 1937 and 1938 issues of the *Greeley Tribune* at the new library before driving to Eaton to have lunch with some Johnson cousins. I got an impressive update on the advances in technology and equipment that were taking place on the farm Farmor and Farfar (Dad's parents) had purchased after immigrating from Sweden.

After lunch, *Mom, Dad,* and I began the five-hundred-mile trek north to Washburn that *Dad* had so wanted to take years ago. As *Mom* receded into the background—maybe to read a magazine or lie on the couch, who knows?—*Dad* and I headed over the Eastern Wyoming plains looking for antelope. *He* especially admired my pair of 12x binoculars that gave *his* truth-seeking spirit a closer look.

That night, I parked Buggles in the Black Hills of South Dakota. *Mom* and *Dad* faded away once I began finding interesting rocks around the RV camp. *They* returned the following afternoon as *we* drove to Washburn after I rode the elevator up the face of Mount Rushmore.

In Washburn, I was allowed to scan the headlines from some of the original 1937 archived issues of the *Leader*—a special treat. Then *Mom* and I went for a long walk. First stop: the war memorial just across the street. I recognized the names of two boys her encompassing spirit had drawn into her choral groups. Although I didn't think to write down their names, this would not have been shared had not some aspect of their serving spirit been internalized in me.

Mom's mindful sadness over their deaths swept over me as *we* headed to the high school *she* wanted me to make sure was still there. On the way, *she* offered up a prayer that those whose names were on the memorial were now singing in heaven. I wondered about which of the spirits of their expressions lived on in the minds of those also walking the streets of Washburn. The steep and now paved three-block hike to the school certainly corroborated Margaret's writing in her May 8 letter, "going up this hill three times a day is hard on leather," another observation of her mindful spirit that now means so much to me.

We headed back down to the river. There, many feet below, lay the smoothly swirling current of the muddy Missouri. I left *Mom* reminiscing with some of *her* fellow teachers while I scooted down to Fort Mandan, the nearby Lewis and Clark winter campsite. Mom wrote in her April 30 letter (*QGW*, 406) that there were a number of

Indians near Washburn and that an Indian school always competed in the county musical contest. Back in 1804, when Lewis and Clark arrived, there were only Indians who welcomed these two strangers and their party of explorers into their villages.

After having dinner in Bismarck with Mary Schneider, Uncle Phillip's daughter, *we* headed east early the next morning and entered Minnesota around noon. Later, *we* passed through some wooded hills and marshy valleys on our way to Clarissa.

Clarissa didn't have a newspaper, but Eagle Bend, just a few miles north and west, did. *We* had hoped to find coverage of Mom and Dad's wedding. Unfortunately, the archived copies were destroyed by a fire a few years back. Fortunately, a few miles away, we found a write-up in the *Lone Prairie Leader*.

Mission accomplished, *we* headed back to Kalamazoo. Along the way, *we* stopped to see Greg Olsen, Aunt Bernice's son, in Minneapolis. Dad's relatives joined the conversation and brought into the present things about their early lives that I had never known or had forgotten. Mom's relatives regaled us when *we* had dinner in Rock Island with a number of my cousins in Uncle John and Aunt Ruby's family. Something similar happened again in Elgin, a suburb of Chicago, when *we* stopped in to see Eldora, Aunt Evodia's daughter.

Chicago was the penultimate stop along the way. In talking with Lois and Anna, my sisters there, *Dad* and *Mom* frequently drew the conversation into a discussion of how their characteristic thoughts and spirits were instilled in our minds and lives.

Both *Mom* and *Dad* receded to the back of my mind on the last leg of my return to Kalamazoo as I concentrated on how to meaningfully share their and my anticipated joy in the hereafter.

CHAPTER 4
I'M ON MY WAY

The following are some of the headlines in the *New York Times* that emerged at the time of the letters in this chapter. June 17: "Germany tightens boycott of Jews. . . . Berlin arrests rise to 1,500." June 19: "Anti-Jewish drive covers all Reich; Arrests mounting." June 21: "Spain's premier charges treason. . . . Rebels continue gains." June 30: "Stocks rise to highest levels since Nov. 1." July 2: "Planes ordered by army. . . . Largest peace-time award includes 13 'Flying Fortresses' and 78 B-18-A bombers."

Back in Greeley, here is some of the news headlined in the *Tribune* at the same time. June 18: "Berlin Jewish shops looted as intensive 'pogrom' is started." June 20: "Damage from Hail and Rain Heavy in Weld." June 23: "Joe Louis flattens Schmeling 2:04," and "Hitler's proud Aryan blood must be at a boil today." June 25: "Jews in U. S. denounced by German Bund [a group of Americans of German descent]." July 1: "Three powers limit guns to 16-inch bore."

In the aforementioned trip, I was accompanied by my parents at various times. Not their physical bodies, of course; they were just shy of thirty when these letters were written. But something had been with me. Had I taken a similar trip with someone I had known who had passed away prior to the Night of the Little Self, I would have thought that something was my memories of them. That all changed once I started seeing the world through the eyes of the little self. I will attempt to describe it in my next song, after Margaret travels to New York with her brother John.

The June 16–July 4 Letters

June 16 [STEVENS POINT, WISCONSIN, JUNE 17, 1:30 P.M.]

Dear Walter,

"Jesus Only!"

I'm on my way to New York with John and Rev. Carl Gustafson & Rev. & Mrs. Ermy Holm. We plan to go as far as Lake Michigan by nine o'clock. We plan to sleep on a steamboat on the lake. We are surely enjoying the trip. It is hard to write when the car is moving.

I received your letter this morning just before leaving. We will be at Mr. & Mrs. J. G. Olander's home at West Oak Hill R. Jamestown, New York, during our stay in New York. Mother and I will not be at the synodical meeting. I hope to see you in a few weeks.

June 16 [LONDON, ONTARIO, JUNE 17, 11:30 P.M.]

Dear Walter,

"All the Way my Savior leads me."

Thus far our trip has gone just fine. We slept on a steamboat last night. We hope to see Niagara Falls this evening. I have never been out of the U. S. before and so I am certainly drinking in the sights. Quite a few ride bicycles and walk instead of riding in cars.

The potatoes are blooming and strawberries are ripe.

We crossed the Detroit river on a steamboat this afternoon.

May God watch between us while we are apart.

In Christ, Margaret

June 19 [JUNE 20, 4:30 P.M.]

My Dear Margaret:

So may the unbelieving world

See how true Christians love;

And glorify our Saviour's grace,

And seek that grace to prove. Hymn # 500

We love, because he first loved us. I John, 4:19.

Margaret, <u>this little sentence certainly explains something that I knew but could not put into words, that is; your love for me</u>. Indeed no love, nor any marriage union could be anything like it should be if it were not true. I hope and pray that our union and our home may always be proof of our Saviour's grace. Within me there is an intense longing for that home and little helpmate, whom I know is hoping and praying for our home and me even now. . . .

Mr. Brown told me they had a terrific hail, this afternoon around Ault. There was some danger of flood on some small creeks, but it is over now. . . .

May God make your trip pleasant and keep you for Christ and me.

In Christ. Walter

June 20 [JAMESTOWN, NEW YORK, JUNE 21, 12:30 P.M.]

Dear Walter,

"Jesus only!"

Mother and I are here at Olanders' but I plan to go to New York City sometime soon. We have had such a nice time travelling. Since we don't know when we are coming home I don't think you should plan to come to Minnesota the week of July 16. I am quite sure that I won't be home then altho maybe the folks will. . . .

Last Sunday we heard the blind pianist and singer Claire Hobart give a concert. It was inspiring. Carl G. sang the liturgy for the morning service. He sang very well. He and John sang a vocal duet also.

The lakes are very beautiful around here. I am enjoying my trip immensely.

Sincerely, Margaret

Margaret didn't feel the need to explain her opening verse. Neither will Walter.

June 22 [JUNE 22, 9:30 A.M.]

My Dear Margaret:

The Teacher is here and calleth thee. John 11: 28

Good morning Margaret: How are you. Did you get seasick or did you only take the boat for a short distance. I thot that you might take the boat almost to New York. A lake excursion on a large boat would be lots of fun and very interesting.

Jamestown is quite a long way north of New York city, is it not? Is it on the Hudson river! It is also the home of one of the largest churches in our synod I believe . . .

May your journey, be safe, pleasant, profitable and may our Father watch over you all the way.

It is about 20 minutes to six. I have set the water and put in the horses. The men who are stacking my hay will harness them. I will finish stacking today if nothing goes wrong or if it does not rain or blow too much. The hay is very good and will be up in nice shape. We have had many very light sprinkles but no heavy rains. Nevertheless there has been much heavy rain and hail all around us.

I will also finish irriggating (I only need one g in irrigating) beans today. They are doing very well. We still have river water. I think we will have plenty of water this year. . . .

I will address this letter to the address you mentioned. When will you be home? Perhaps for the Sunday following this one.

Now may he watch over us always and haste the day when we may see each other again.

Yours in Christ, Walter

June 26, 1938 [JUNE 26, 9:00 P.M.]

My Dear Margaret:

"Verily I say unto you, What things soever ye shall bind on earth shall be bound in heaven; and what things soever ye shall loose on earth shall be loosed in Heaven." Matt. 18:18

This statement of Christ's, seems to me to place a great responsibility upon us. For it seems to say; that if we are to be bound to God and his Son in heaven we must be bound to him on earth. We are also reminded again in this Chap., that our sins are forgiven to the extent that we forgive others their sins against us.

This verse would be hard for us to read and take if it were not for the two verses which follow it. ["Again I say to you, if two of you agree on earth about anything they ask, it will be done for them by my Father in heaven. For where two or three are gathered in my name, there am I in the midst of them."] For only thru praying for and receiving God's grace and help can we hope to become bound to Christ. . . .

When will you be back to Minn.? I have received no answer to my proposition, but I suppose that is because the letter was addressed to Minn. This house would need a little work on it and I think it would be nice if you could come out here and suggest how you would like it. I have not spoken to [the] Browns but I believe they would be willing at least to buy paint, paper and kalsomine.

We seldom have water any later than the 2nd week in Sept. The fall work will begin the first days in Oct. so if we are to be married in Sept it would be best to set the date somewhere around the 18th. Even so, it would be far better if we could have the house ready for us even to the extent that we could buy the furniture and decide where we would have it. . . .

I do not want you to have to come here and have a lot of hard work to do. The fall work requires long hours on my part and I could not help then. There are adjustments to be made. Farm life is a glorious life (I think) but it is different than any other. There is much physical labor for the men and plenty of work for you too. But I like to work and especially farm work. I do not worry much about the adjustments tho for I know you were quite interested in farming and you love nature and Gods handiwork. To get up in the morning with the smell of the cedar sage and pine in your nostrils always make me appreciative of the glorious handiwork of our Creator and heavenly Father. . . .

May we recognize; and I ask your help in doing so; that all things are God's and that we are his stewards, put here to care for and increase all that he puts us in charge of. I am praying for the Grace and will power to tithe from our very beginning. To give generously to the cause of his kingdom and to spend carefully for our necessities and nothing on foolishness or needless things.

I love you very dearly. I feel a little lonely for you this Sunday afternoon and sincerely hope that it be God's will that we be united soon and need never be apart for a very long time. I received your cards and thank you for them, yet I hope that I receive a letter soon. . . .

I finished stacking hay Wed. afternoon. It was put up without rain on it and is pretty good. I think there is about 40 tons this cutting. I get 1/2 of the hay and 1/3 of all the other crops. I have had quite a bit of help the last two weeks. Altogether it has cost me so far about 75 dollars for extra labor. . . .

I do not know what I will do the fourth of July. Frank Lunn is back at Cheyenne and wants to go fishing. But it costs me quite a bit to go fishing in Wyoming and I am trying to get along on as little money as possible. Maybe Ill stay at home and work. To me it is no particular holiday. No one stops to thank God for our independence, or at best only a few do.

They only shoot fire crackers, get drunk, dance, go to shows and raise cane so to speak. . . .

Frank Lunn said that Dr. Pihlblad is recovering his mind now. I guess he has been pretty blank for a long time. May God help him physically and mentally and if he can again serve Bethany may he do so as God would have him do it.

Now as I close I would reach my hand across the miles and as I clasp yours in mine, tho only in spirit, yet it is a warm and sincere spirit that asks God to bless you and keep you; for his kingdom and for me as a helpmeet to walk through this earthly life and then when we are called may we and ours be found on the solid rock on the other side.

With much Christian love, Walter

I had a lot to learn before *glorious* was a word I associated with farmwork. There was much I enjoyed about it, but it was financially distressing and the work, unrelenting. My daydreams were filled with other possibilities.

Mom and Dad laid the groundwork for my pursuing those possibilities. Mom gave me her love of music and Dad his love of reasoning. Whether at work or play, they helped with whatever I needed to know. They encouraged me to read. The complete sets of the *Book of Knowledge Encyclopedia* and *Encyclopedia Americana* beckoned in our living room.

It was just a matter of time before the first engaging possibility—other than Major League Baseball—would come my way. It happened when Mr. Hoff, my grade school principal, organized a trip to a science demonstration put on by the Dupont Company at the Ault High School auditorium. I heard a train stereophonically approach on the left, roar by, and fade off on the right. I saw a rubber ball shatter when thrown to the floor after being frozen solid. I was hooked. From then on, I wanted something to do with the scientific understanding of nature.

My interest in science evolved over time. As a senior in college, I was wondering what to do with my bachelor's degree in zoology when my brother Paul mentioned his wanting a master's degree in biostatistics. It seemed a perfect fit for me as well. I had enjoyed geometry in high school and calculus in college and was fascinated by quantifying the reliability of data-based conclusions in my first college statistics course. I ended up spending over thirty years doing statistical analyses of scientific data with cell biologists, neural scientists, veterinarians, chemists, pharmacologists, and computational chemists in preclinical pharmaceutical research.

June 27 [WILMINGTON, DELEWARE, JUNE 27, 11:00 P.M.]

Dear Walter,

"My Church! My Church! I love her ancient name!"

Today we have been to the [tricentennial (?)] celebration. We witnessed the ordination in Bethlehem

Church in N. Y. City yesterday. Tomorrow we plan to spend in Philadelphia.

I received your letter while I was in Jamestown. We are not planning to return via Jamestown. I hope you are almost thru with your haying now.

It has been raining very much. We have had very good time together.

When I see you I will tell you about the trip.

In Christ, Margaret

June 28 [JUNE 29, 4:30 P.M.]

My Dear Margaret:

"For the Son of Man also came not to be ministered unto, but to minister, and to give his life a ransom for many." Mark: 10:45

Margaret, it seems such a long time since I heard from you that I find myself fearing that you are not well. My hope and prayer is that you are instead enjoying your trip and benefiting from it.

I have often wondered about that which we call love. How is it expressed? As we look about us we certainly see many different ideas of how it should be expressed. But I think <u>Jesus makes it very plain here that we are to express our love thru service to one another</u>. Certainly, He who was love, knew best how to express it.

I hope and pray that soon we may serve and help each other and may it always be to the glory of God. . . .

Have you seen fifth avenue, the Brooklyn Bridge and the Empire State Building? Perhaps you have also seen the Statue of Liberty. The true statue of liberty is the cross, for there we are liberated from sin, the worst slave master of the world.

Yours in Christian love, Walter.

P.S. Last Sunday I came to our prayer hour at 6 P.M., but that meant 9 P.M. there.

Should you change your mind, about staying as long as you had planned, please let me know immediately. I must either come see you soon after the 4th or wait until the latter part of July or 1st of August. I do wish you could come out here for reasons that I have already mentioned. Tho I long to see you, do not cut your trip short. I think I can stand it for awhile yet, if I only get a letter once in awhile. Please write as often as possible.

July 1 [GARY, INDIANA, JULY 2, 1:00 P.M.]

Dear Walter,

I received you letters when we returned to Jamestown last night. You seem to be getting along fine with your work. Thanks for the letters.

I am not able to answer your question yet but I can let you know later. We are returning home sooner than planned.

The Lake Erie is just beautiful. We are just stopping to get some ice cream.

In Christ, Margaret

July 4 [EAGLE BEND, MINNESOTA, JULY 4, 6:00 P.M.]
My dearest Walter,

"Jehovah will keep thy going out and thy coming in From this time fourth and for evermore." Ps. 121:8.

We returned safely from our long trip last night. It seems like we are just spending a day or so and then we are to move on again but that is a thing of the past.

Walter, you were wondering if I were ill. I have been well on the entire trip but there has been so much to do and when one has been travelling the whole day and ~~get~~ gotten to retire at 11:30 and get to sleep only until 6:30 one cannot write letters. I didn't get your letters until I returned to Jamestown and then I received one this morning. I never knew where I would be and therefore I couldn't give you any other address than to Olanders.

I surely thank you for the invitation to your place. Won't it be fun to be together in our future home? It will be fun to be together looking for furniture and to pick out wallpaper....

The trip was interesting. We didn't stay as long as we had planned. I did get to see quite a few unusual things. When I see you maybe I will tell you of them but they will keep until we are settled in our home. Some winter night after a day's work I can tell of them.

You mentioned that you thought sometime in September would be a good time for the wedding. You're busy the first part of the month. I think you have been awfully busy and by the way your letters sound you have been waiting for letters. I maybe should have stayed at home but I know you are glad that I could take a trip.

Walter, when do you want me to come? It won't take me long to get ready. . . .

I wish you were coming here but I will be glad to come since you mentioned so many reasons. I will enjoy being with my dear Walter.

With Christian love, Margaret

I don't recall Mom ever complaining about giving up her chances to travel because she married Dad. She was always somehow in the present.

July 4 [JULY 5, 4:30 P.M.]

My Dear Margaret:

"Righteousness exalteth a nation;

But sin is a reproach to any people." Prov. 14:34

With millions of dollars being spent upon whiskey, beer, soft drinks, firecrackers, shows, rodeos, dances and only our Father in Heaven knows the other numerous ways in which hard earned money is spent today. Why? To celebrate the anniversary of the winning of Independence for the United States of America.

300 or more years ago a little band of Pilgrims, braved the Atlantic Ocean in a frail wooden boat, in order to find and settle in a land where they might worship their God and Father as they chose and saw fit.

Today—300 years later we commemorate the anniversary of the winning of our independence from England; which made sure our privilege of religious freedom. How do we do it? Serving the devil with a beer bottle in one hand and a firecracker in the other. Of course it is easy to conceive, (if you have a strong imagination) of even half of our people, taking so much as 3 minutes to get upon their knees and thanking God for his bountiful grace and goodness. I wish I could believe one out of 25 did this, but I cannot. Our 4th of July celebration is certainly a farce.

But enough of this, I have much to thank God for and I am thankful for so many things. First of all for you. I have been very lonesome for you today, and hope to see you soon. I do not know when you will get this or read it, in fact I don't know where you are at. I will await your decision as to whether you will or will not like to come out to Colorado in order to help plan for our home. I do not know where to address this letter, so I will send it to Clarissa. . . .

I shall close for a while. . . .

Good morning Margaret:
"Jesus Only"

Another beautiful day has been given us. We had a very nice shower last night, so it is cool and fresh. We have not had much warm weather lately or at anytime for that matter. We have certainly had a wonderful spring and summer. If it does not hail or fall down too badly, we will have the best grain crop we have had in many years. Our grain is mostly barley. All the other crops are doing fine also. . . .

It is noon and I must go to dinner. May God keep you. . . .

In Christ, Walter

The Fourth was a holiday for our family. We usually went to Greeley to watch the parade in the morning and to picnic in the park in the afternoon. A few times, we would stay for the fireworks show on the fairgrounds even though we all knew some of the evening chores would go undone.

We occasionally lit fireworks at home, nothing impressive and always a disappointment when one fizzled out. We were not allowed firecrackers for safety reasons even though Dad's father always awakened the household on the Fourth with a firecracker blast.

A Glimpse Back

June 19. How would you explain Margaret's love for Walter? Or Martha's love for me? Or the love of your soulmate or best friend?

June 26. Walter believes that "we are to express our love thru service to one another." How might serving heal divisions within our families, neighborhoods, communities, and countries? Who might be most helpfully served? Why? How?

June 28. Though Walter longs to see Margaret, he doesn't want her to cut her trip short. What are some of the pros and cons that Margaret could read into this message?

July 4. Walter raises the issue of religious freedom. What types of freedoms are most important to you? Think of a time you felt truly free from restrictive rules and pressures. What understandings and experiences were involved?

Watching the Embodied and Expressed Spirits at Play in These Letters

In my extended reflection on Margaret's remark, "In a way you were present too," following her May 17 letter, I briefly noted that she could tell her colleagues about Walter because aspects of his spirit resided materially in her mind. By the same token, Walter could share with Margaret what "Jesus makes . . . very plain" in his June 28 letter because aspects of Jesus's spirit resided materially in his mind. Several issues arise in comparing the significance of these two capabilities once you distinguish a person's embodied spirit from their expressed spirit.

Your embodied spirit is responsible for those physical expressions that others will perceive as directly attributable to you. Your direct expressions are conveyed in the words you choose, the gestures you make, the actions you take, the things you make, and the possessions you acquire.

Your embodied spirit comes into being at your birth, pervades your body, and perishes upon your death. For the most part, your direct expressions are even more transient. Unless recorded, your voicings die out quicker than echoes; unless filmed, your movements fade even faster.

The consequences of your expressions are entirely different. These are vast, enduring, and endlessly complex. They may be obvious and seemingly transient as in a tit for tat. They are deep and durable when lives and worlds are changed. Aphorisms, such as "your reputation precedes you" and "what goes around, comes around," naturally arise.

The material changes your direct expressions bring about in the minds of others I like to call your expressed spirit. The direct expressions of your embodied spirit cease upon your death. Your expressed spirit somehow lives on in the minds and expressions of others.

Margaret's letters are illustrative of this. Prior to her correspondence with Walter, her Washburn colleagues in North Dakota had had no awareness of either Walter's embodied or expressed spirit. Margaret was learning all about Walter's embodied spirit in his letters, but she closeted these letters and kept their disclosures to herself. When sent the ring, she put it on her finger for all to see. This unleashed Walter's expressed spirit in Washburn. In her April 18 letter in the preceding volume, Margaret wrote,

> This diamond has called for a lot of explanation. So many have wanted to see it and have asked about you. The teachers are so filled with questions that I almost wished you could have been here to answer some of them.

Although Walter's embodied spirit wasn't at the Washburn school, his expressed spirit certainly was. It set the topic of a dinner conversation with her school colleagues—at least, that is what we might surmise from what she wrote in her May 8 letter:

> This evening at supper the teachers asked me if sometime while they are travelling in Colorado if they could come to visit us. They surely love to talk about what farm-life is really like. This evening they said for me to be very careful when I am out gathering eggs so that I don't reach in the nest and touch a snake. There aren't very many snakes on your place usually, are there? . . . I believe that they use their imaginations when they get good and started in teasing.

Here, we see the play of some embodied and expressed spirits organizing around the guiding spirits they choose to heed. In the first sentence, we see a grouping spirit bringing the teachers together over dinner. A deceiving spirit, so often a part of the repartee of teasing, appears in the third sentence. It may have elicited chuckles around the table reflecting Margaret's discomfit, all conveyed in her fourth sentence. A forgiving spirit comes through in her last sentence.

On reading Margaret's words, bits of the expressed spirits of her colleagues entered Walter's mind. The spirit of deception that came through in the teasing, as harmless as it might have seemed to Margaret's colleagues, irritated Walter. That spirit undermines the spirit of truth Walter has been heeding in his attempts to increase Margaret's awareness of farm life and its joy. In his May 12 letter, he responds, "First we'll discuss snakes some. Tell those who tease you, that 99 out of every 100 snake stories are exaggerated to the point where they cannot be believed. Just plain baloney!"

Truth triumphed in this case. In the first of her two May 17 letters, Margaret wrote, "The subject of snakes has not been brought up since I wrote to you but now I have my answer ready."

Now consider the encounter Margaret describes in her second letter:

> Mrs. Jefferis was up this evening. She told me last night about what hard financial difficulties they were in for about four years of their first

years of marriage. She said she was glad that they hadn't waited to be married. Walter, I haven't told her anything about finances except last spring when you wrote that you were coming. I told her of your plans and then I received word that you couldn't, so I had to tell her that your plans didn't materialize because your stock hadn't sold as you had anticipated.

So much is at play in what Margaret shared of the expressed spirit of Mrs. Jefferis. A spirit of service is evident in her taking the time to talk to Margaret. Had Mrs. Jefferis not felt a spirit of Margaret's worthiness, she would not have made the effort. A spirit of truth underlies the aptness of what she said. Her awareness that something had been troubling Margaret and her wanting to know why stemmed from a spirit of mindfulness. Margaret was struck by Mrs. Jefferis's concern for the marital decision facing her and Walter. Strangely, Walter was not. He didn't comment on what Margaret had shared of the expressed spirit of Mrs. Jefferis.

Contrast that to his response to what Margaret shared of the expressed spirit of Jesus when, without further comment, she opened her June 6 letter with the verse from Mark 10:45: "The Son of Man came not to be ministered unto, but to minister, and to give his life a ransom for many." Three weeks later, Walter opens his letter with that same verse and then writes,

> I have often wondered about that which we call love. How is it expressed? As we look about us we certainly see many different ideas of how it should be expressed. But I think Jesus makes it very plain here that we are to express our love thru service to one another. Certainly, He who was love, knew best how to express it.

These are two different reactions to two different expressions of love by two different people, neither of whom Walter has met.

Truly thanking someone for a gift is an expression of love in which so many of our guiding spirits are involved. Margaret could have, and probably did, directly thank Mrs. Jefferis for her thoughtfulness. We can't know. We weren't there, and Margaret didn't say. However, thankfulness for a person's act of kindness is also conveyed by furthering that person's expressed spirit. Margaret did that. A spirit of service is embodied in Margaret's devoting time to organize and carefully script what she shared. A sense of Mrs. Jefferis's worthiness comes through in what she wrote, and what she wrote is pervaded by a spirit of truth. None of this would have happened had Margaret not been mindful of the kindness of Mrs. Jefferis's act.

Walter could not directly thank Jesus; the embodied spirit of Jesus had perished centuries before. Instead, he shared with Margaret the serving spirit of the love of Jesus, thus furthering the *expressed* spirit of Jesus.

CHAPTER 5
GOD'S HARD WORKING

Here are some of the headlines in the *New York Times* that cropped up during this period of Walter and Margaret's correspondence. July 5: "France occupies Isles off Hainan; Tokyo perturbed" and "Gettysburg holds greatest Fourth. . . . All branches of army perform for Blue and Gray and fireworks light sky." July 9: "Arabs in Palestine begin wide strike as violence grows."

On July 5, the *Greeley Tribune* reported, "513 dead in July 4 celebration," and on July 16, "American-Jewish homesteaders appeal for aid after repulsing armed Arabs."

The July 5–July 17 Letters

July 5 [JULY 7, 6:00 P.M.]

My dear Walter,

"Delight thyself also in the Lord:

And He shall give thee the desires

Of thine heart." Ps. 37:4

The above verse, I know, has been and is very precious to you and to me. This morning I awakened at 4:00 o'clock. I just kept thinking about you until I finally got up to pray about you and about our future. Then I just became so happy in the realization that you love me and that you are an answer to prayer that I just couldn't resist writing to you again altho I sent you a letter yesterday. . . .

Walter, I am so glad that you are well so that you can enjoy working. . . . May God give you strength to continue and may I help so that you may not be hindered in any way by my being with you. <u>May God lead both of us so that the kingdom may come by us</u> also as we pray in one of Luther's petitions of the Lord's prayer. . . .

In Christ, Margaret

Wednesday night.

"I was glad when they said unto me,

Let us go unto the house of Jehovah." Ps. 122:1. . . .

John told me, just before he was to leave for Eagle Bend this evening, that I could probably go to Mount Carmel for next week's session of Bible study. The Eagle Bend & Clarissa Leagues have rented housekeeping cabins and there is room for at least one more girl. I was at L.B.I. [Lutheran Bible Institute] camp two summers ago and I surely received much benefit, spiritual and physical. . . .

Today mother and I baked bread, cinnamon rolls, and buns. . . . Ebba said, "The bread tastes very good." Again, I wish you were here to have a slice of bread and butter.

Walter, when we were at Jamestown with [the] Olanders, who formerly were members of my father's congregation in Tennessee. They gave me a nice breast-pin, a pickle-dish, a handkerchief, and a five dollar bill. This sentence is so twisted I can't straighten it out so I had better go to bed. . . .

Should we get married sometime in September there are but a little over two months left before we are one. May God keep us ever His and may that union draw our hearts nearer to Him always. He can supply all our needs as He sees fit.

May god richly bless you. Good night & sleep well.

Yours in Him, Margaret

Thursday morning.

"There's a wideness in God's mercy

Like the wideness of the sea.

"There's a kindness in God's mercy

Which is more than liberty."

This is one of the stanzas which I memorized this morning while I mended some hose. It is a wonderful song. We sang it a few times on our trip.

I received your letter this morning. Thank you. You express my thoughts concerning our nation's observance of the Fourth of July. Mother and I spent the day at Johns. . . .

Mother is sewing a rag rug for us. She said that she would help me get started to make a quilt. I don't know where she plans to get a hold of the material for the quilt but I know that she knows. . . .

Yours in Christ, Margaret

In writing "that the kingdom may come by us," Margaret is referring to the first phrase of the Lord's Prayer in Matthew 6:10: "Thy kingdom come, thy will be done on earth as it is in Heaven." As the daughter of a Swedish Lutheran pastor, she knows Luther's Small Catechism, in which Martin Luther writes, "The kingdom of God comes indeed without our prayer, of itself; but we pray in this petition that it may come unto us also."

It is difficult to imagine how the kingdom in heaven coming to Earth might be experienced. In the Gospel of Thomas (verse 113 in the Thomas Lambdin translation), Jesus is asked, "When will the kingdom come?" He replies,

> It will not come by waiting for it. It will not be a matter of saying "here it is" or "there it is." Rather, the kingdom of the father is spread out upon the earth, and men do not see it.

An analogy may suggest the difficulty our prophets face in trying to describe what is underway that we do not see. *The Free Dictionary* says a hydra is "any of several small freshwater polyps . . . having a naked cylindrical body and an oral opening surrounded by tentacles." Like most plants, it consists of many highly differentiated cells. Under appropriate conditions, the cells of a hydra can be gently teased apart. Forced to go their separate ways, the cells return to their undifferentiated amoeba-like forms. Brought back together under the right conditions, the cells reconfigure and resume their complementary forms and, thereby, resurrect another hydra.

When unconstrained in their amoeba-like state, they go their separate ways, possibly thinking, *Life is a struggle and I'm going to get what I can and whenever I can.* Such thoughts might be anthropomorphically attributed to metastasizing cancer cells. When suitably constrained, something of an entirely different order must take place for a newly organized hydra to emerge. Some of the hydra cells may, in essence, be sharing the thought, *If we help each other find our respective callings, something will happen in which everyone will have role to play based on their particular situations, understandings, and talents.*

As we struggle for possessions, we sometimes speak or hear others talk of the rewards of cooperation, the joy in finding one's talents, and the freedom and fulfillment in going about one's calling. We often use metaphors and analogies. Speaking of a time to come, Isaiah (65:22) says, "They shall not build and another inhabit; they shall not plant and another eat; for like the days of a tree shall the days of my people be, and my chosen shall long enjoy the work of their hands." In Luke 13:19, Jesus says the kingdom of God "is like a grain of mustard seed which a man took and sowed in his garden; and it grew and became a tree, and the birds of the air made nests in its branches." We cannot know fully what life will be like once the spirit of the realm of which they speak engulfs humankind. After watching the movie *The Road Back*, Margaret expresses the enigma holding us at bay when she writes in her December 4 letter (*EG*, 73), "I wish that greed and hatred were not in the human heart. Maybe then, however, we wouldn't long so for heaven."

July 8 [JULY 9, 4:30 P.M.]

My Dearest Margaret:

"for all things are possible with God." Mark 10:27.

Margaret, I certainly was glad to receive your letter and to hear that you are well and have enjoyed your trip. I am certainly glad you could go and I am sorry if I in any way caused you to cut your trip short. True I was anxious for a letter. I would like to get one every day, but I know that is not necessary and no more possible than it would be for me to write that often.

It will be fun to have you come out and so much fun to plan with and for you. May our prayer be, that it all be unto the Glory of God and his kingdom.

Now the question is when to come. I am going to give you an outline of League, S. S. & Church activities which may help you to decide. I will be quite busy on the farm the greater part of this month, meaning that I will have to work the greater part of the daylight hours. There will be a little spare time in the latter part of the month. Next week; 15, 16, 17 of July, Greeley League is host to the Colorado district L. L. July 31 is S. S. Picnic. First week in Aug. is Bible Camp in Estes Park. Then in between our regular League, S. S. and Church meetings. . . .

I shall send a ticket immediately, which will make it possible for you to come at your most convenient time. If you could take at least two weeks, it would be nice if you could come sometime in the last week of July and stay until sometime in the 2nd week of Aug. I don't think I can spend much time at least not all week at Bible camp, but you could if you would like. You have been there so you know how beautiful it is up there. (this 2 wk period) It is also in haying and threshing time and that might be interesting.

It would be nice if you could stay longer and we would have time to take a few short trips to the mts. But this you will have to decide. I believe we can get much done in two wks. Decide how the house should be decorated, (In the meantime you can study catalogs and pictures of rooms, and then when you come we can study them together) and furnished. We can also see how we are at stretching the Colorado Cart wheels or the almighty dollar, as we shop for what we need. If barley is a good price 50¢ a bu. or better I may sell some and get in some money that way. If not I will borrow what I need until I thresh the beans sometime in Sept. I just thot of something, I want you to remember. Whatever we do, I hope we can and will discuss it and plan it together and I know we can and will do this; but <u>the house is your workshop</u>, where you necessarily have to spend most of your time, therefore; as much as possible and means will allow, we want it as you want it. That is the way I want it and will like it.

It is pleasant to think of planning and working with you, but it will be such fun when my Margaret is really and truly right here. I cant help it. I must say it. You're the nicest, sweetest person in the world. and I hope that I will not be a disappointment to you. But I do not fear this much, because I know that in Christ, you have the power to overlook much and I too shall always come to him for help and direction. This is our greatest duty and obligation and it will always be our greatest source of happiness and peace. May we always be found faithful. . . .

In Christ, Walter.

July 9 [JULY 9, 3:00 P.M.]

Dear Walter,

. . .

Mother has promised to sew my wedding dress but she says we must help her. I want it to be of white color, don't you? Mother wants to have a white dress to wear this summer maybe she can get some good material and then plan to wear it at the wedding. I want her to walk down the aisle with me.

. . .

Walter, I am so glad that you mentioned that you wondered where we would be spending our vacation next year. I was of the opinion that one seldom gets a vacation when living on a farm. It will be fun to plan a trip. I can hardly wait to see and be with you. I pray for our future and it seems the more I pray the more I long for you. May God grant that we may always come to His throne of grace and then we need not fear that our love will grow cold. I believe that He is the One who keeps the homes of Christians warm with love, don't you? . . .

    Yours in Christ, Margaret

July 12 [JULY 12, registered]
    My Dear Margaret:
    And hereby we know that we know him, if we keep his commandments. 1 John 2:3
    A new thot, or rather a clearer understanding of a certain question which has been in my mind became clear as I looked upon this verse and the others around it. For lack of time (this is 2 P.M. Tues. July 12) I shall not try to explain it. Perhaps it will come to us at sometime and I can also get your opinion. *Margaret I look forward to the day when we may discuss, read and learn together*. Two viewpoints are valuable in any study, don't you think so. . . .

    In answer to a statement & question in your last letter, I must say; that I certainly do believe that it is God, who keeps any love warm. The source of love is not from within us, but from God. If we stray to[o] far from that Holy fire we soon become cold. Tho I have never known the family Altar at my home, I am determined that you and I shall begin that practice from the very first day. I ~~could~~ can only do this if you, too, are determined to do the same thing. Therefore I thank God for the knowledge that, this is your desire. I know it is. I am convinced that if we omit this at all that we are building upon sands that are ever shifting. I am glad, because I have the knowledge that you pray for me, for us and for ours. It greatly strengthens me. . . .

I am sending 25 dollars, which will cover most of the traveling expenses to Colorado. When you come I will pay any additional expenses. This only gets a 1 way ticket but I still plan to take you back, God willing.

Our plans are not always the best so circumstances come up that make it impossible to follow them. This must be God's hard working don't you think so? Right now I have a horse that has sleeping sickness. It is a serious sickness in horses and quite often fatal. I hope I do not lose him but *if I do I am asking grace to accept it as I should.* . . .

In Christian love, to my Margaret. Walter

July 13 [ALEXANDRIA, JULY 13, 11:00 A.M.]

My dear Walter,

"The law of truth was in his mouth, and unrighteousness was not found in his lips: he walked with me in peace and uprightness, and turned many from iniquity.["] Malachi 2:6. . . .

In yesterday's class we studied about how the priesthood had fallen away and misled the people but that God still loved them. Then God expresses in the first verse which I quoted what an ideal priest does; speaks, and acts. He does turn many from iniquity. . . .

Walter, there is a young married couple that are spending a while here at camp. They live about 180 miles from here. They are farmers. I had quite a talk with her but not for as long a time as I should have liked. She said that she just loved to live on the farm. They left their one and a half year old boy at her mother's place. I was thinking sometime maybe you and I could attend a Bible Camp. That is a wonderful vacation and rest at the same time that one can receive spiritual nourishment.

Ida just arose and is busy peeling potatoes. <u>She said that I should continue writing</u>. It seems whenever I have something extra to do I always get help. Last night three of us girls from Eagle Bend and Clarissa sang the trio "Drifting" and the night before I sang "Under His Wings" at the evening services. The girls helped me to get ready. . . .

Yours in Christ, Margaret.

July 14 [JULY 15, 4:30 P.M.]

My Dearest Margaret:

"He is not here; for he is risen, even as he said."

Matt 28;6.

How important that statement is. <u>If it was not in the Bible or was not true, this world would be vastly different</u>, so much so that the most vivid imagination could get no picture of it. . . .

I am tired and must soon go to bed, but I wanted to send just a little letter to you. Soon I shall see you and we may talk together and plan together.

I lost the horse that was sick. He was suffering so much that I shot him this morning. <u>It was quite a loss but I think it will all work out for the best</u>. Tomorrow I get water again for 24 hours. . . .

How are you enjoying Bible camp? Is it very hot or does the lake help to cool it? How many people do you and Ida have to cook for? It will be good training, for the time when you will have a hungry farmer to cook for.

It is cool here again after a fine rain yesterday afternoon. It was quite warm before that though.

Tomorrow night, our convention begins and continues thru Sunday. Frank Lunn and Arnold Anderson (Rev Erlander's brother) will come down to stay with me. I wish you were here, but you will probably attend the dedication at Clarissa. . . .

In Christ, Walter

"It was quite a loss but I think it will all work out for the best." That was Dad. No complaints but the "grace to accept it" per his request of God. Dad skillfully navigated time's river in a way that brought joy to his life. He occasionally struggled with the river, but he didn't fight it; he embraced it. He seemed to appreciate seeing how the twists and turns eventually resolved themselves.

Shooting that horse was much more than a personal loss. Dad always spoke affectionately of the willingness of his horses to work and their enjoyment of a good rubdown at the end of a hard day. Dad and Mom moved twice before settling on the farm where I grew up. By then, Dad was farming with tractors. Years later, when I left the farm, the horse harnesses were still hanging in the barn.

It was only a couple of years after my two older brothers and I entered college that Dad too left the farm. Mr. Wells, our landlord, had said Dad could lease the farm for as long as he wanted. When Mr. Wells passed away, the farm passed to an heir who felt our neighbor could farm it more profitably.

Changing careers after turning fifty was one of the sharper bends Dad had to negotiate. Although he had worked nights at the Great Western Sugar Beet Factory for the needed money, factory jobs didn't interest him, however steady the hours and good the pay. Although he liked farming and being around farmers, his pride and independence ruled out his being a hired hand. By the time he put his farm machinery up for sale, he was enrolled in real estate classes and naturally felt his familiarity with water rights, knowledge of the land, and long interest and experience in farming would be valued qualifications when working with buyers and sellers of farmland.

July 16 [JULY 16, 6:00 P.M.]

My dearest Walter,

"Let me more of their beauty see

Wonderful words of Life."

Walter, my heart is brimming over with joy. I just reached home and here I found a letter from you waiting for me. I had felt sort of sad, too, that I had been at camp all week and hadn't received a single line from you. I told mother that I couldn't understand why I hadn't received a letter from you the whole time that I was at Mt. Carmel. She said I had gotten two letters and that she had sent one letter to me in the tennis shoes which she sent. I didn't wear my shoes a single time so that is why I didn't get to read a line that you had written.

Thank you so much for [the] registered letter containing the check and the pictures. . . . I don't know when I shall start on the journey to you but I shall let you know. . . .

In Christ, Margaret

July [17–18] [JULY 19, 6:00 P.M.]

My dearest Walter,

"Bless the Lord, O my soul, and forget not all His benefits: Ps. 103:2.

Thanks for the letter which I received yesterday. . . .

I am sorry that your horse died. I hope it isn't the one that was gentle. I hope none of the other horses get the sickness. <u>The death of the horse, I'm afraid, means quite a loss for you. May God supply you with another horse.</u>

Walter, there isn't much time left before I'll be on my way to Colorado. On the 29th of July is mother's birthday and in the evening is an ice cream social. John wants me to sing a solo that evening. I don't like to travel on Sunday as I guess it will mean that I start out on Monday morning the 31st. It will surely be fun to be with you in our little home for a little while before coming there to stay a long long time. . . .

This letter is short but I must get it mailed on this afternoon mail.

Yours in Him, Margaret

The disappointing sale of his farm machinery when having to leave the farm was a much bigger loss. In talking with Dad after the auction, I could feel his disappointment when he said only a few showed up because of a heavy morning rain, so rare in sunny Colorado. It was not easy to hear what he had expected his tractors, cultivators, harvesters, and other farm machinery and tools to bring followed by what they brought. Yet he never complained nor looked back. He, like Mom, lived in an eternal now. The conversation soon turned from the disappointing sale to the interesting things he was learning in his classes.

Fortunately a few years earlier, Mom had renewed her teaching certificate at what is now the University of Northern Colorado. By the time they left the farm, she was teaching the third grade in Windsor. With only a grade school teacher's salary and four children still living at home, Dad's first few years as a real estate agent were a tough go. Yet Mom excitedly told me about each of his early listings even after it became apparent that the wanted sales commissions were often not forthcoming. Still, she backed his decision not to work on Sundays.

One afternoon, while Dad drove me around to see some of his farm listings, we stopped at the entrance to one of them. He noted how he had hung up his office's FOR SALE sign at the entrance and how the farm had been advertised. He then related how surprised he was to see that the FOR SALE sign had been taken down when he drove by a couple of weeks later. On being asked why, the farmer said he had been offered a cash deal if he would reduce the price by the realty commission. The farmer apologetically offered Dad a thousand dollars for his part in the deal. Dad said he wanted a check made out to the realty office for the full commission or he didn't want to be paid anything. Not anything was what he got—financially. Spiritually, he retained the freedom to live life the way he enjoyed it, open and honest.

A Glimpse Back

July 5. Margaret sees a kingdom she would like to help bring about in her home and world. What are you trying to bring about in your home and world? What guides your efforts?

July 8. "The house is your workshop, where you will spend most of your time, therefore; as much as possible and means will allow, we want it as you want it. That is the way I want it and will like it." In what distinct ways do the understandings in each phrase in this statement come into play when evaluating the practical health of your working relationships?

July 12. Walter is looking forward to the day when he and Margaret "may discuss, read and learn together." How do you weigh such things among all the others you would like to do with someone with whom you live?

July 13. Ida, a friend at camp, feels Margaret should keep writing rather than help prepare the upcoming meal. How do you think Ida weighs the various short-term and long-term benefits of Margaret's writing, and how is Ida's spirit materially a part of those benefits?

July 14. Walter feels that if the resurrection of Jesus was "not in the Bible or was not true, this world would be vastly different." What different issues of truth and fact do you see in his statement, and how do you weigh their relevance to your beliefs and those of others?

CHAPTER 6

A LITTLE SWELLING ON MY FOOT

Here are some of the headlines in the *New York Times* that were hot topics of discussion during the time of the letters that follow. July 15: "Roosevelt offers to join a world disarmament move; He sees trend to 'Disaster.'" July 16: "10 Arabs are slain; Jerusalem tense." July 17: "Reich seizes farms owned by French. . . . Troops occupy 2,500 acres." July 22: "Japanese threat to oust 'invaders' defied by Soviet." July 24: "British ask Czechs to strain a point in placating Nazis." July 28: "Franco considers plebiscite in rebel Spain; resents report army keeps people in line." July 30: "Russians and Japanese in fight on border; invasion repulsed with losses, Soviet says." July 31: "Mussolini defies Vatican warning in racist dispute." August 3: "Soviet hurls six divisions and 30 tanks into battle with Japanese on border."

And here are some headlines from the *Greeley Tribune* during the same time. July 21: "Hitler assures Great Britain he desires peaceful solutions of the outstanding questions." August 10: "Maneuvers of Germany's army stir new fears in Europe that outbreak of war looms." August 12: "1,300,000 are called to colors in Germany; French watch closely."

The July 19–July 29 Letters

July 19 [JULY 19, 4:30 P.M.]

My Dearest Margaret:

Keep thy heart with all diligence;

For out of it are the issues of life. Prov. 4:23

We have a beautiful, cool, and sunny morning with which to start the day. It rained beautifully yesterday afternoon. It was not a heavy rain but a very beneficial shower. We have certainly been blessed with an ideal summer for our crops and I would ask you to thank God with me, for them. For if it be God's will we will share their fruits. How much fun it will be to share it with one so dear and one who is able and willing to recognize and thank the one who creates and gives all things. I have always said I would never marry unless I could and would be given a Christian girl and now that I have known you and how much that means, I know it could never be any other way. Dr. Kendall said at our convention that if we settle the question of our relationship to God and his Son our Lord first of all, that countless thousands of other questions that come up will be easy to meet. . . .

But, I must hurry on to other things. The time you mentioned in which you planned to leave for Colorado is probably as good as any. It looks now as tho it will be a busy time for me so I would rather see you postpone it for a week in order that I might have more time with you. On the farm, (I guess that's true in anything), you have to do some of these things as they come up. Perhaps the next week will also present plenty to do but as a rule our busy season ends after the threshing and 2nd cutting of hay is up. But whatever time you come will be alright, perhaps it would be fun and just as well to see farm life as it is. I like the hurry and the work. So whenever you feel it is right and God's will, then come. I am anxious to see you as soon as possible and hope and pray that the visit will be interesting and enjoyable.

I am sending a program of our convention. I have no time now to write more concerning it. (When I get in a hurry I'm some scribbler.) . . .

In Christ, Walter

P.S. I'll put in one clear sentence at least. Whatever time you come will be the right time.

July 19 [JULY 21, 3:00 P.M.]

Dearest Walter,

"Therefore I say unto you 'All things whatsoever ye pray and ask for, believe that [ye] receive them, and ye shall have them!" Mark 11:24

I am so happy tonight so I decided to share it with you even tho I haven't any ink to write with. I plan to buy a bottle of ink tomorrow morning. . . .

I wrote a letter to Evodia telling about our plans of being together for about two weeks.

Walter, lets not spend much money on furniture. We can get a long with less than to get deep into debt. I do hope that your remaining horses will keep healthy and that your crops will yield bountifully and that you will receive a good remuneration for your work and all that goes with it. . . .

I just wonder what you are doing this evening. Maybe you are doing just the same as I, writing a letter. Soon we will get a chance to speak instead of write. . . .

Lovingly yours in Christ, Margaret.

Wednesday evening. [July 20]

"Draw nigh unto God and He will draw nigh unto you." James 4.

Now I am lying propt up in bed with one of my feet raised up on a pillow. I thought I would tell you about what happened today before I go to sleep. . . .

Now I am finally coming to what I have planned to write to you about for almost a week but I wanted to know more about it first. While at camp I noticed a little swelling on my foot, I asked the Mt. Carmel nurse about it and later a deaconess, Sister Minnie, about it and they both said it was a tumor and for me to see a doctor about it but not to worry about the swelling.

Since Margaret Setterman asked Ebba & I to ride along to Bertha we got a free ride to the doctor. I told him about the tumor. He said that because of the place of the tumor he thought it best to have it cut away . . . In the afternoon I had the operation. I was conscious during the entire time. He asked me if I knew how to say the word "ouch." Later he asked me maybe I could say it best in the Scandinavian language. Miss Thiel, the assistant nurse said in that language it is "oj, oj, oj." . . . He said that I should do very little walking for a month.

Now—how about my going to Colorado? I want to come but it will be very hard to walk around shopping when I am to be so careful. I am glad to be over having that swelling which I thought was but a mosquito bite. I surely would like to see you and talk to you but you're so far away. Perhaps I could have waited until after I had returned but I am quite sure then there will be so very much to do and I would have to remain in bed in the midst of all the work. I surely thank God for the health he has given me. May He guide us in all our ways.

I am sending you a picture that was taken in New Brooklyn. . . . The picture flatters me so I guess that is the reason I chose to send it to you. I wish that I could come to you instead of the picture but I guess we will have to wait for a while to see one another.

May God grant that this operation will draw me nearer to him.

Yours in Christ, Margaret

I'm unfamiliar with the spelling "oj, oj, oj" but often heard something like *oy, yoy, yoy* accompanying someone's personal pain or disappointment at Swedish gatherings.

July 22 [JULY 22, 4:30 P.M.]

    My dearest Margaret:

    . . .

    Good morning Margaret; . . . I received your last letter yesterday. It told me many things. It will be just fine to have you come in the 1st week of Aug. That is not a long time in the future so it will not be long until I shall see and speak to my darling.

    I like to read of the work that you do in the way of sewing, crocheting and other fancy work. I am also glad that you are able to be with and help Maria Olson. God often hides his jewels in the rough and if someone does not help to polish them and bring them out, their talents would be lost to the world. Even God's own Son had a herald and forerunner in John the Baptist. Tho it is not exactly the same thing, <u>a little encouragement and proof of love on the part of a friend, may help to bring out some wonderful talents in another.</u> . . .

    I believe we can arrange sometime to go to Bible Camp at Estes Park and then during recreation time you and I can climb upon some mount and thank God for the blessing he has given us.

    Now as I close it is with a prayer that god will ever keep you and bless you.

    In Christ, Walter

July 22 [JULY 25, 6:00 P.M.]

    Dear Walter,

    . . .

    This morning I finished crocheting an edging for one of my pillow cases. This afternoon I plan to start its mate. Mother said she is going to give us a pair of pillows. I seldom use a pillow when I sleep but it looks better to have two on the bed during the day.

    Tomorrow I am to go to the doctor to have my wound cared for. I have been in bed since Wednesday afternoon. There are so many things that I should like to do but I do want to obey the doctor so that I can get well soon so instead I lie in bed. . . .

    Mother brought in a skirt of a yellow dress for which she wants me to sew a blouse. I said that I would sew one. I am a very poor seamstress but I am confident that she will help me. . . .

    P.S. . . . Rebecca didn't see Isaac until she came to his home to be married. I have seen you at Colorado and in Minnesota so I can wait until you come to Minn. for the wedding.

July 23 [JULY 25, 6:00 P.M.]

    Dear Walter,

"And all flesh shall know that I, Jehovah, am thy Savior, and thy Redeemer, the Mighty One of Israel." Isaiah 49:26c.

This is one of the verses which impressed me when I read the forty-ninth chapter for devotions. This is a good quotation to use in worshiping God. I hope we can memorize a large number of verses to have in our minds and hearts to use especially in our prayers. Rev. A. B. Anderson said, "Pack your prayers with Bible passages."

Walter, you mentioned that it seemed that you could read God's handwriting in God's taking away one of your horses. I am afraid that the losing of that horse meant quite a loss financially. I hope that your other three horses will remain well to work for you very faithfully.

During the past few days I have been in bed according to the doctor's orders. . . . Dr. Will took off the two clamps and dressed my wound again and told me to return next Monday for another dressing. He doesn't want me to walk much on my foot for at least three more weeks. <u>Now I have additional time for meditation; God knew my need</u>.

Walter, since I can't be around much I believe it best not to plan a trip to Colorado until I come with you as Mrs. Walter Johnson. What do you think about it? These are eight weeks left until the eighteenth of September, the date which you wanted the wedding. That time will go so fast I have a lot to do before so I can certainly keep busy. . . .

Since I have been in bed and have realized more and more that I had better give up the trip to you, your folks, Evodia & Reuben, and our home to be, I have thought that I would love to have the twenty-five dollar check go toward or for a piano. It seems a piano is such a friendly instrument to have in a home. We would have so much fun playing the piano. I would like to teach you to play hymns during the winter while you have a chance to rest from field work. I think Evodia could pick out a piano that is inexpensive and yet has a good tone. Then when we get company we can sing songs. Well, Walter, I know you want a piano maybe even more than I do and you know when the best time is to buy one so if you think best not to use the check for that it will be all right. Maybe you have something to use it for now so just go ahead because I won't be cashing the check.

"Again Thy glorious sun doth rise
I thank Thee oh my god."

Good morning, Walter. I can imagine that you have been up quite a number of hours. I hope you are feeling fine and eager to work. I feel fine and wish that I could be up and working. . . .

Walter, only eight more weeks! Mother and I have been trying to figure out where we can find sleeping quarters for the wedding guests. I think it is going to work out just fine to get places. Don't you think the wedding should be in the evening! All of the weddings that we have had have been at that time. We can go to church for morning services together.

I would like to go to Colorado before the wedding but since the time is so short and mother wants me here and I can't be on my foot very much and since we are not flushed with money also the 50th anniversary of the church is to be celebrated next month I will have to stay home. . . .

In Christ, Margaret

P.S. The little girls are playing "Ring Around the Rosy" in front of the window. They changed the game and now it is "Drop the Handkerchief." They keep changing the games; now it is "Hop-Scotch." *"De tala Swenska alkt jamt."*

Although Walter saw "God's hard working" in the death of his horse, Margaret didn't attribute that death to God. She does, however, find a reason for God's bringing about her tumor. She ends her letter with a Swedish phrase Walter would understand: "They speak Swedish very well."

July 24 [JULY 24, 10:00 P.M.]

My Dearest Margaret:

"God is our refuge and strength,

A very present help in trouble" Psalm 46;1.

I am so sorry that you must suffer some pain now, as well as the anguish of being unable to be about and carry out your plans and work. But one thing I want to assure you. Tho

I long to see you, it does not bother me in the least that you could not come. I am afraid that you might think I would not understand and be too disappointed. Please do not worry at all. I do understand and I think that you used excellent judgment in doing as you did. It is a very good idea to nip these troubles in the bud so to speak. . . .

  I wish this afternoon I could sit with you or near you. I would like to hold your hands at least and if you could I should like to have you sit in my lap. I could also read to you and help you to pass the day. This is impossible, but I can and do pray for your quick and speedy recovery.

  As for coming to Colorado. You need not worry about that. All things work for the good of them that love God. Perhaps you can come in about 3 wks. You need not walk much. I will also watch my chances to come if only for a short visit, altho I would like you to have your ideas carried out in the house. This will work later and I could not work on the house for quite a while anyway.

  Since you are not worried I think I can keep from worrying too. You look so well on the picture; Evodia [Margaret's sister, who was married to Reuben, Walter's pastor] saw it last nite and the first thing she said was She certainly doesn't look sick there does she? . . .

  I am so glad that you would rather get along with a little, than go deeply in debt, just to have a lot of new furniture and equipment. As I said before, I always hoped to have some means when I got married, but I think love is by far the most important; especially love with Christ for the foundation. <u>It seems that God did not intend for me to get these means.</u> Perhaps; so that we might strengthen our faith and launch out on him. Today I would rather have you by my side, and we together in Christ than anything or everything else on earth. God will provide for us and ours as he sees fit. Of course we must work but that will be fun with you to come home to, and I hope and pray that you will like it too.

You said that you hoped I would always like to come to our home. *If you only knew how I long for the day when I can come in to our home and the dearest little lady in the world.* When we can see the sun rise and set together, study nature, the crops, our animals, nature's life, of plants and animals and God's Holy word together. Again I say I cannot thank God enough for providing me with a helpmate who can see and will worship God in all nature and life. For most certainly all things are of God and point to him; to those who would seek him and see him. Even the wind or breeze which just sprang up tell of his might, power and love. . . .

Now may he who starts the wind, rolls the clouds, stills the waves, creates new life and keeps the old, all to his glory; make you well soon, as well as keep you and bless you unto a life of service in his kingdom. Then when he wills it, may I too work by your side in this glorious service.

In Christ, Walter.

. . .

I feel very well. I lost less weight this summer than ever before even tho I work all the time. I also got a new suit last nite in preparation for our big day. I bot it then because of an anniversary sale at our best clothing store: 25% discount meant quite a bit and I had to have the suit.

This letter captures how I came to see Dad's freedom in his relationship to God, his love for Mom, and his joy in farming.

July 25 [JULY 27, 4:30 P.M.]
    My Dearest Margaret:

. . .

I wish that I might see you today. I would like to know how you feel and how your foot feels now. Was the tumor on the bottom of the foot or to one side? I always wish and especially at a time like this that we were not so far apart, so that I could help you pass some of the time away. I imagine that you are sewing and crocheting part of the time and that will help to pass the time.

I sent a little box of candy to you. I wish I could place some flowers in your room. I always like them and I think they help to make things more cheery and pleasant. I am somewhat curious to know how the railroad and postal workers handle this package. I did not insure it and tho I asked them to they did not mark it as being fragile, so I wonder how it will look when it arrives. . . .

I shall write some more later, but now I bid you goodnight and leave you to the care and keeping of our gracious Father in Heaven.

5:53 a.m. Wed.

Good morning Margaret:

I (I had to fill my pen) wonder how my little lady's foot is this morning. Has it bothered you much or kept you from sleeping? I am still hoping that it will be mended and strong again soon so that you can come to Colorado in the near future. I would like to be told to do this or that with this ceiling and walls or that woodwork, by the Dearest Person in all the world. But if for any reason you cannot come then I will make arrangements to come at the earliest possible chance, which would not conveniently be before the middle of Aug. or perhaps the 20th. . . .

Mr. Brown has been sober for a long time so I wish you could meet him at such a time. <u>Tho he is not a Christian, in fact does not believe in anything that he cannot see or feel, yet he is a nice man</u>. Even when he is drunk he is not loud or noisy, tho somewhat talkative. If our patience can hold out we may be able to help him to understand that the most real thing in the world is our Father in Heaven. He has no basis for what he believes. He only believes that way to appease his conscience. . . .

I have not lost any more horses. I had them vaccinated against sleeping sickness so there should not be much danger of them getting it. . . .

Perhaps you will progress better than the doctor thot and be able to be there too. How I do wish you could, but God's will be done, and may he keep you always.

In Christ, Walter

Walter feels that Mr. Brown believes only in entities that can be seen and felt. Certainly, Mr. Brown has seen the same "earth and all its beauty" that evokes Walter's thoughts of God. Many of Walter's metaphoric takes on that Earth can be traced to the Bible. No doubt, Mr. Brown's life views are also traceable to one or more influential books as well. Yet neither Walter nor Mr. Brown sees a basis for what the other believes.

What should we do when we question the basis for each other's perspectives? Three possibilities are easily imagined: We can "go on the attack" to either win over or wipe out the opposing view. We can disengage by adhering to authoritative schools of thought. Or we can "live and let live" while seeking common ground and trusting in everyone's desire for joy and fellow feeling.

July 27 [JULY 28, 6:00 P.M.]
Dear Walter,

"Jehovah is good to all:

And His tender mercies are over all His works." Ps. 145: 9.

Let us remember that God showers His tender mercies on all His created beings. Think how wonderful God is [to] show His goodness to all mankind. . . .

Thanks for the letter and candy. Walter they made me lonesome for you. I wish you were here to share the candy and I could sit in your lap meanwhile with my one foot that is still under special care on the bed. I have tried being on my foot for a short while during the day but I have to [go] back to bed because my foot begins to swell a little. I guess the doctor is right about the time to recovery. The candy will help the time to go faster. Thank you. . . .

To think that you have gotten your wedding suit! I wish that I could see it but I can wait until the wedding. I don't plan to show my dress until that day either. I just wonder what mine will look like. It is fun to plan, I think. I wish God would whisper in my ear what the wedding will be like because even tho we haven't made our plans He already knows what it will be like. . . .

John just came so he'll take my letter.

Yours in Him, Margaret

Margaret airmails the following letter:

July 29 [JULY 30, 6:00 P.M.]

    My dearest Walter,

    "I am my beloved's:

    And His desire is toward me." S. of S. 7:10.

    This evening I just reread the book, "Love That Lasts" by Samuel Miller. How happy I am that you truly love me and what makes me more joyful is that God loves us both and His desire is toward us. John took Mrs. Martin and me to our respective doctors this afternoon. The doctor looked at my wound and said, "Well, this time we won't bandage it." . . . God kept His promise "All things work together for good to those who love the Lord." Walter, I do love altho my heart is often cold. He warms it up when I come to Him in reading His word and thru prayer and especially thru singing. There are so many things which I long to relate to my dear Walter but if you knew what time it is you would send me to bed so I shall continue this letter in the morning, God willing.

    May God bless and keep us and teach us to know His will. I thank Thee, God, for being so gracious unto me for healing my wounded foot so speedily. May I walk where Thou dost lead.

    Good night, Walter

Saturday

    "Praise God from whom all blessings flow." . . .

Since my foot has healed so fast I plan to leave for Colorado on Monday morning. The bus leaves here at 10:30 a.m. I don't plan to stop off any place except to change buses. If all goes well and as planned, I will get to Greeley on Tuesday evening. I presume you know when the bus reaches Greeley. It surely will be fun to see you. I hope that all will go well. I'll soon get to see our little home. Oh, but I'm looking forward to coming. I'll get to meet your mother and sisters and to see your father and Bruce again. We can see Evodia, Reuben, and Grace at Estes Park if you continue as you once planned. May God bless our visit together and with you and my dear ones.

Yours in Christ, Margaret

A Glimpse Back

July 22. Walter suggests that "a little encouragement and proof of love" may help others develop their talents. What expressions of others helped you find and develop your talents?

July 23. Margaret claims God knew she needed more time for meditation. How might God have communicated that needed time to her?

July 24. Walter is looking forward to embarking on married life with Margaret on a farm but lacks the means he had hoped to have when doing so. How do you see the factors underlying his optimism? What would be the most important factors for you if you were in his shoes?

July 25. Even though Walter is a Christian who doesn't drink and Mr. Brown is an atheist who occasionally drinks too much, they seem to have decided to "live and let live." What approaches work best for you when working with others who hold different spiritual perspectives?

CHAPTER 7
YOU STILL LOVE ME THO, DON'T YOU?

The following are some of the headlines from the *New York Times* during the period of time of Walter and Margaret's next round of letters. August 16: "Hitler opens vast army exercises testing machinery of mobilization." August 28: "Hitler warned by Britons that an attack on Czechs may result in world war." September 2: "Hitler bars a Czech truce. . . . France to raise army to 825,000." September 6: "Nazi insurrection put down in Chile." September 11: "Goering pledges Germans to protect the Sudetens [ethnic Germans in Czech lands]." September 15: "Americans informally told to return home; many ship bookings to Europe canceled."

Here are some headlines in the *Greeley Tribune* Walter may have read prior to traveling to Clarissa, Minnesota. August 30: "French and English move to save peace." September 6: "Hitler defies world blockade, silent about Czechs." September 7: "Hitler advises rejection of Czech offer."

Roughly a year later, on September 1, 1939, a *New York Times* headline would read, "German army attacks Poland; cities bombed; port blockaded; Danzig is accepted into Reich." More than two years after that, on December 8, 1941, another would read, "Japan wars on U. S. and Britain; makes sudden attack on Hawaii; heavy fighting at sea reported."

Margaret and Walter's courtship ends with them heading to the altar in Clarissa. We will be able to join them, but to enter into the experience, we must turn from the ticking now of the big self and embrace the eternal present of the little self. That inversion of time frames is described in the last song I will sing here, "Joy in This Life and the Life Hereafter," the title of which is what I was unconsciously seeking when, in my teens, I asked Dad, "Why are you a Christian?" (*EG*, 2).

The August 15–September 10 Letters

August 15 [AXTELL, NEBRASKA, AUGUST 16, 2:00 P.M.]

My dear Walter,

"Thou wilt keep him in perfect peace whose mind is stayed on thee."

Reuben sang the song that contains the above verse. The song continues to ring in my mind's ear so I want to share them with you. . . .

This evening our program consisted of two solos by Reuben, the solo that I have already mentioned, a duet by Evodia and Reuben, a trio by us three girls and then the songs I sang were "I Bow My Forehead to the Dust" and "I Heard the Voice of Jesus Say." The chapel [Bethphage Mission in Axtell, Nebraska] is such a beautiful one. . . .

Today you are to cut your bean crop. I hope that you may harvest a fine crop. I wish that I could bring you lunch with Bernice and Mrs. Johnson. Soon I can bring you some all by myself just so Tom [one of the workhorses] doesn't scare me away.

Greet your folks. I will be looking for some of them up in Minnesota in a few weeks. Did you get to speak to Mr. Brown about putting in places to plug in lamps etc. I presume some work has been done in the house that will soon be our home, god willing.

May God bless & keep you, Walter.

In Christ, Margaret

Margaret's quoted verse is a spiritual anchor that rings in her ear because of the understanding, rhythm, and choice of words. All minds eventually acquire spiritual anchors of one sort or another. Walter and Margaret are attracted to those they find in the Bible. Here is the complete verse from Isaiah 26:3 of the King James Version, along with seven other translations:

"Thou wilt keep him in perfect peace, whose mind is stayed on thee: because he trusteth in thee" (King James Version).

"The confident mind You guard in safety, in safety because it trusts in you" (Jewish Study Bible).

"Those of steadfast mind you keep in peace—in peace because they trust in you" (New Revised Standard Version).

"Lord, you keep those of firm purpose untroubled because of their trust in you" (Revised English Bible).

"A nation of firm purpose you keep in peace; in peace, for it trusts in you" (New American Bible).

"This is the plan decreed: you will guarantee peace, the peace entrusted to you" (New Jerusalem Bible).

"You will keep in perfect peace those whose minds are steadfast, because they trust in you" (New International Version).

"A nation of firm purpose you keep in peace and security because it trusts in you" (The Inclusive Bible: The First Egalitarian Translation).

The understandings in these translations are arguably quite similar. Multiple translations often give us new insights and move us closer to the original thought, but often at the risk of losing the actual words, rhythm, and ring that struck us initially.

August 16 [SAINT PETER, MINNESOTA, AUGUST 17, 12:00 P.M.]

Union Bus Depot, Omaha, Nebraska
My dear Walter,

. . .

When we came to Dorchester we learned that the bus doesn't stop in town so Reuben took me along to Crete. From there I took the train to Omaha. I took a taxi to this depot. The train fare came to $1.50 taxi fare to 20¢, and now bus fare home amounts to $8.50. . . .

Did you know that one can buy stationary paper and envelopes for a penny each at a bus depot? That's what I paid for this. It certainly helped me out as I didn't have any paper with me and I have no postal cards.

While at Lincoln I ate at the train depot restaurant. My dinner came to 35¢ and I got such delicious food that I can't see how they could serve it for that price.

Only one hour longer of waiting and then I sail away on the bus. I wish that you were here. We could talk together about our future days. By this time I suppose you have finished cutting one-half of your bean crop. I hope the yield is what you expected or even more. No doubt Mr. Brown has been busy talking to you about the house. . . .

Fairmont, Minnesota

Good morning, Walter.

I hope that [you] slept as well as I did. When I awakened this morning, it was 5:30. Several [people] told me that I had certainly slept for a long time. The bus driver said that I awakened one time and asked "Was that Fairmont?" He surely liked to tease me about it. . . .

There is a little boy on the bus who is travelling from Calif. to Eveleth, Minn. all alone. He has his name and address on a piece of paper that is sewed on to his suit coat.

The driver of the bus had charge of his money & sees to it that he gets along fine. He calls him "cow puncher."

St. Peter, Minnesota. Now I am here at Aunt Mary's. She was certainly surprised to see me. Lillie and she have had new electrical fixtures put in. They have bought a Monarch wood, coal, & electric combination range. It is a beautiful stove & takes very little fuel. . . .

Yours in Christ, Margaret

Aug. 16 [AUGUST 17, 4:30 P.M.]

My Dearest Margaret:

. . .

I did not know that I should miss you so much. Even tho I am busy yet I cannot keep away that lonely feeling.

Did you have a nice trip and arrive home safely? I hope you are well as you go to work on the wedding plans. I am certainly glad that it will only be 32 days until we are united in marriage as well as in Christ. . . .

I am quite sleepy. I had to be with the water a great deal last nite and since I had so little sleep the nite before I need to get to bed. I can only cut beans in the mornings while the dew is on so I must arise at 4 and get going while I do.

I am sending the window dimensions. . . .

In Christ, Walter

The beans could only be cut early in the morning. The bean cutter sheared off the plant at the root. Once the sun evaporated the dew, the pods quickly dried, and when dry, many of the pods snapped open when shaken by the bean cutter. Only those beans still in the pod would be picked up and shaken loose by the combine later in the heat of the day. When the bin of the combine was full, the harvested beans would be augered into a truck and taken to market.

Aug 18 [AUGUST 19, 6:30 P.M.]

My Dear Margaret:

Search me, O God, and know my heart;

Try me and know my thots;

And see if there be any wicked way in me,

And lead me in the way everlasting. Psalm 139;23,24

<u>If all the way our Savior leads us, we certainly do not have much to ask beside, do we</u>?

Soon you are home again, very likely tonight. Then to work. Say; I wonder if yellow curtains for the kitchen would help to brighten it especially with very few red spots.

I scrubbed the front room and dining room this evening and tomorrow the floor of these two rooms is to be varnished. The painter will probably be thru Sat so I suppose I can begin cleaning it up then. That Walltex certainly looks nice in the kitchen. So does the wood work. I wish you could see it and the other rooms. I have not asked for any wall plugs. Mr. Brown is on another drunk so that it is hard to talk to him.

As we move in and settle, some of these little problems, we will solve the light problem with them.

If we find we can get along with Mr. Brown we can even do some of these things ourselves, next spring. Right now I do not care to ask for too much.

We are (my class) planning to go to the mts (Aug 25–27). I believe at least 20 will go. Please remember this unto God, for I would like to make it something good and worthwhile. . . .

I must go now and wash wood work so that the painter can finish Sat. . . . Yesterday as I said, I scrubbed two of the rooms. Perhaps I should have said nothing of this ability. Oh well, it will be fun to help you, but the waiting for the opportunity is hard.

I finished cutting beans today. They are good and the price is fair . . .

In Christ, Walter.

The next letter brings a foretaste of some logical whirlwinds that would later spice up life on the farm with Margaret.

August 19 [AUGUST 20, 3:00 P.M.]
My dearest Walter,
"He leadeth (us) me O blessed thought
O words with heavenly comfort fraught

Whate'er I do, where'er I be

Still 'tis God's hand that leadeth me."

Walter, this song is meant for both of us. Now I know this song by memory. Maybe you have memorized it also. We can sing it together when we are riding in the Black Hills.

I wish that you were here tonight. I am so lonesome for you. I just wish that I could rest my head on your shoulder.

. . .

Walter, Ruby just showed me her wedding ring. It has hers & John's initials engraved in it such as R.M.M. to J.N.L. then the date, then Song of Solomon 8:6. [Set me as a seal upon your heart, as a seal upon your arm; for love is strong as death, jealousy is cruel as the grave. Its flashes are flashes of fire, a most vehement flame.] All this was abbreviated. John has the same except the reverse order with regard to the initials. Papa and mama had the same in their wedding rings. I was wondering if we could also. Maybe it would cost too much tho I wish that I could talk to you right now. I didn't know that I would miss you so much. . . .

I stopped off in Minneapolis to enquire about bridal apparel. Everything seemed so high but I didn't go to the cheaper stores for information. Maybe it will be that I can sew my dress with about 96% mother's help and many good suggestions from the girls but if mother's strength does not seem to be able to endure it I am sure that if I try with less help of hers and more from the girls and then do as much as I can I will have a pretty nice dress that would cost only $5.00.

. . .

Walter, while on my trip home I have been thinking about our house. I am very well satisfied with it all except for that dark brown oil cloth ? that looked like wood in the kitchen. I have thought it best, if by any means possible, to change that to a very light blue or green or ivory. I am quite sure that large border on the walls from the mop board up to about four feet will make the rooms look darker than we should like when we live in the rooms. May be that you have thought the very same thing. No doubt your mother and sisters have also. Would there be any possibility to change the color? If I remember rightly, we didn't send for the Armstrong rugs so that we can easily change that to another color. I do like ivory colored walls for the kitchen. What do you think we can do about it. I am sorry that I didn't realize it while I was down at the Moffat shop. If it can't be changed by any means without causing ill feeling, it will be all right. I have been thinking about my being in the kitchen when the furnishing[s] are darker and it makes me feel like wishing it changed.

Walter, I have so many things to tell you now and I can't understand why I didn't think of them while I was with you. I wish that I could see the nicely painted bed room. I hope to start sewing or making curtains. Did you succeed in getting new plugs for putting in lamps and so forth.

<u>If we get a very good kitchen set that is durable and nice looking, a good range and a comfortable bed that looks nice too and a piano I'll tell you we should live like a king and a queen, don't you think.</u> . . .

It is 12:20 p.m. or should I say a.m. I have forgotten but I expect you to know. Walter, you'll have to help me many times. You recall you helped me spell knives correctly.

Walter, I do so long to see you and sit in your lap. I am glad that the color of my hair my height and voice pass your approval altho I can't do selecting of furnishings and other things as I should like. You said that it would work out all right.

May God bless and keep you always. If He continues to shower His benediction on our courtship we will soon be united. . . .

Good morning, Walter. This morning I made breakfast. It didn't work out so well, I guess except for the coffee. John said that the coffee was very good. Thanks for the letter[;] it came the same afternoon as I came home.

In Christ, Margaret

Aug. 22 [AUGUST 22, 9:00 A.M.]

My dearest Margaret:

. . .

Another day is over and now before I retire I want to write a few lines to my Margaret. I did not know I would miss you so much. It was so much fun to have you out here where I could see you and be with you, that I failed to estimate rightly, how hard it would be to pass the time when you are not here.

. . .

Today my right knee has swelled up a great deal and is quite painful. I do not understand it. I injured it a few years ago and it may be that I aggravated it when I crawled on the floor while painting. I have to irrigate again tomorrow and Thurs. We leave for the mts, so I hope it gets well quickly.

The house is finished except for a few jobs that I will do, so there is now no chance for any changes You will be surprised how well the kitchen colors blend and how bright it is. Light curtains and Linoleum will make it very pleasant.

I was going to order the linoleums yesterday but when I saw the ones for the kitchen I could not make up my mind. The one for the dining room I could not get in the size we wanted so we will have to pick another. I was going to let it go until you came but today I decided I better get at least the ones for the kitchen so that I could put it down and have the stove up (Browns stove). I believe now I will ask God's help and direction and get them for both rooms for we will have plenty to do any way. I think it will all work out all right don't you. I shall pick colors that will light up the rooms and harmonize pretty well with the new decorations. Then we would only have to pick out a rug for the front room, a throw rug for the bed room and the dining and front room furniture. When you come we can decide about the bedroom. We are welcome to Browns bed for a while if we want it. . . .

Now I say goodnight to my darling; always asking God to keep you and bless you.

In Christ, Walter.

August 22 [AUGUST 22, 6:00 P.M.]

Dearest Walter,

"There is therefore now no condemnation to them that are in Christ Jesus." Romans 8:1.

Walter, hasn't Christ done much for us and then right now there is no condemnation if we are in Christ? May he be all in all [everything] in our home life and everywhere.

This morning I received you[r] letter. Thank you. I am glad to hear that the painter has succeeded so well. So the Walltex looks very nice! I am glad it does because I have thought it over and the more I think of it the more I am in favor of having the kitchen and dining rooms combined. That kitchen is large enough to have it serve very nicely as a dining room. We could have a large enough front room then. It would be nice to have that smaller room be as a bed-room. The Walltex which is somewhat darker than that should be for a kitchen will add to making it seem like a very well-suited room for dining in. . . .

Do you recall the time we had an evening dinner with a couple who had been married a short time before Christmas? Mother happened to sit beside her, Mrs. Swanberg, at the dinner. She asked mother if I was planning to get married on the 11th of next month. "O no," Mother said "it is to be on the 18th." So now it has been told. I don't see why we should keep it secret any longer up here, do you? I am so glad about the fact that the day is soon drawing near. . . .

Next week Ruth and I are planning to go to Minneapolis to buy my wedding dress and that which I will need for the great day. They told me in the cities that at quite a number of the fall weddings the bride is wearing a suit with no veil (which would surely look out of place) but with a corsage on my left shoulder that is divided in two by a ribbon bow, it will all look nice. The sentence isn't clear but I think you understand it anyway.

The suit would not be white of course. This lady that spoke to me is one of three or five who have their special work of selling and planning bridal apparel. Would you prefer having me wear a long white dress with a veil? She said that since the wedding is to be so late it is best not to wear white. . . .

What could we do about Mr. Brown? Ever so often I think of his sad condition. I do remember him in prayer but when Liquor is legal it makes the problem much more difficult but nothing is too hard for God. "Prayer changes things."

We will all remember your group that is . . . to study and rest in the Word up in the mountains.

Yours in Christ, Margaret

Aug. 23 [AUGUST 24, 4:30 P.M.]

My Dear Margaret:

"He guideth me in the paths of righteousness, for his name's sake." 23rd Psalm.

I just finished dinner and just before dinner we finished our little house. That is the painting and varnishing. I shall wash the windows and screens when ever I have time. Then after I thresh and get some money I will get the dishes, utensils and linoleums also. I wish you could see the house. The kitchen is actually beautiful, so are the other rooms for that matter. I have to put one more coat of varnish in the bedroom and one more coat of paint on the kitchen floor.

How I do miss my little Margaret. . . . This month will be longer than the period from Christmas until your visit. But it won't be too terribly long until I will see my little darling again and then I shall not have to leave you again except for short periods once in awhile. . . .

Good morning; to my Dear Margaret:

. . .

It is such a beautiful morning. I am so happy and have so much to be thankful for that I can almost be certain that the sun is shining upon you also and I can almost hear you sing as you prepare breakfast and prepare to go to church.

I went up to Browns to do the chores, so I took another peek at our little house and it certainly looks clean and bright now. I believe you must have a good sense of taste and of color combinations for the colors certainly blend and harmonize.

I must go to breakfast and shave and dress for church, so <u>I pray that the</u>

<u>Lord may bless you and keep you.</u>

The Lord make his face shine upon you, and be gracious unto you:

The Lord lift up his countenance upon you, and give you peace,

In the name of [the] Father and of the Son and of the Holy Spirit.

In Christ and with much love, Walter.

We often said this benediction after meals instead of the Lord's Prayer.

August 25 [AUGUST 26, 3:00 P.M.]

Dearest Walter,

"Commit thy way unto the Lord: trust also in Him; and He shall bring it to pass." Ps. 37:5.

This is really the first chance I have had to write to you. . . .

I am grateful that I can fulfill my promise about letting you know in my next letter regarding the length of time that you need to be here prior to the wedding. . . . The Clerk of the Court told me that you don't need to come at any set time and that I could make the application alone but this application must be at least five days before the wedding day. I want you to come as soon as you can nevertheless. . . .

A week from next Sunday the Mora Church will be ready for services. Philip and Edna have asked me to sing a solo. I wonder what song I will be singing. <u>God knows but I shall have to do some searching to know which one it is</u>.

Ebba just came up to tell me that I have only seven minutes left before this letter must be in the post office. I have so many things to relate so I shall just wait until tomorrow. . . .

Now I have returned from evening services. . . . Pastor Lundeen . . . said that sometime he wants to visit in our home. Won't it be fun to have him? I do think it would be fine to have two bedrooms. Then we can invite pastors to stay at our home during mission meetings. We want our home to be a true trustee's home. . . .

When the Luther and Bethany college choirs gave a concert at the church, we always had one or two of the members spend the night. The concerts, overnight stays, and morning breakfasts were special treats for Mom and Dad.

To think that our home is so near being ready! I wish that we could be able to furnish it quite well after two years have elapsed without going into debt. Rev. Lundeen says that he would much prefer going without than to go into debt. He said also that if one starts going into debt one has a hard time to get out and sometimes one never gets out. He also said that he doesn't owe anyone any money at present if I recall rightly.

Now I must retire. Mother, who is my sleeping partner, has already gone to bed and I know that she wants me to go to sleep soon. May God bless and keep you, my Walter. I do long to see you.

Shall we have the wedding on the 25th. I prefer having it a week later than the 18th. That date is Bernard's and Judith's wedding anniversary also.

Yours in Christ, Margaret

Aug. 28 [AUGUST 28, 3:30 P.M.]

My Dearest Margaret:

"I can do all things in him that strengtheneth me."
Phil 4:13

You sent to me in your last letter the verse from Romans 8:1. It is a wonderful statement, and I say thank you for it and then send this verse to you as a sequence to it, for as he took away our condemnation he also supplies strength to do all things which are according to his will. . . .

My knee is almost well again. In fact, I took the boys on a 15-mile hike one day and felt no ill effects. I did not even get stiff or a sore muscle.

You and I shall spend some of our time in these particular mts in the years to come for I like to look up to the mts from whence our help comes. It is the place we went through last year and saw the beaver working.

Now as I close this short letter with a promise of another soon, I want to do so with a prayer for you, and I do always pray for you and love you more each day for it.

I miss you so very much, but I can bear it because he gives me strength and also the knowledge that we shall soon pray and be strengthened together carries me along very well.

<u>My knee was quite bad but God took care of it quickly and nicely, just as he did with your foot. So I am sure that he will care for us.</u>

Good morning to my darling.

In Christ, Walter

When Dad and Mom last visited my home in Michigan, he was hobbled by Parkinson's disease. Pictures of him striding across a farm field would flit through my mind as he patiently struggled to take the next step with me holding his arm. I don't know how many times Dad said to me when I was young, "Don't shuffle your feet!" Now he struggled to get one leg to move ahead of the other, occasionally remarking something to the effect of, "It's strange; my legs know where they are supposed to go, but they don't go." A few years later, he lost his ability to talk but didn't lose his ability to greet you with a welcoming smile.

Before they started on their twelve-hundred-mile return trip after that final visit, we gathered together in a ring for a last moment together holding hands. Dad prayed for "travel mercies." I silently prayed that the drivers they would meet along the way would be alert for the unexpected.

I backed the car out of the driveway for Mom. She shouldn't have been driving, but with Dad telling her the roads to take, she got on and off the interstates, made it through Omaha, and arrived safely at their home in Greeley.

Aug. 28 [AUGUST 29, 5:30 P.M.]

My Dear Margaret:

"For the Sabbath was made for man and not man for the Sabbath." Gospel of Mk.

I am quite sure that this verse comes from the 2nd Chap. 7th [the twenty-seventh] verse but not sure enough to say definitely. . . .

This verse was the Key verse of our Bible study this morning and I came to see a clearer meaning of it than ever before. <u>I had never thot of it as a gift from God or as a very special privilege of man's, to have this day of rest as a day to rest and worship and recuperate after six days of work.</u> Had you? God did not make the Sabbath just to pin man down to a certain rule. No; God made the Sabbath for man and gave it to him to use as he chooses for the choice is pretty much our own. Yet many people, in a frantic quest for pleasure and satisfaction of a certain longing; use it in such a way as to leave them worn and tired and without hope of ever satisfying that inward longing.

Dad extended this thought regarding the Sabbath to all God's commandments. They were not rules to pin you down. They were guidelines for enjoying a fuller and richer life.

I came home a few moments ago from the water, in the hope that I would receive a letter from you and I did. It is so good to receive your letters. When a period of over a day elapses I begin to long for them and for you. I cant understand this lonesome feeling and why I miss you so much for I am very busy, but I just cant seem to think of other things as I should. I think I better get married as soon as possible so that you can be near always and help me with my work.

About the wedding date, I would say, set it at the day you like best. There are two reasons why I think the 18th is a good time at least not much later. One is I want to see you as soon as possible and 2nd, if we wait until the 25th we will have to hurry home, if we are to straighten up before we start the potato and beet harvest which will begin the 3 or 4th of Oct. . . .

The more I think of the question concerning the date of the wedding, the more I think it should be the 18th. The harvest time comes in rapidly and I doubt that the folks could come much later. As for morning or evening that would not matter and I certainly think we should change the plans so that Phillip could be there. With a simple wedding it should not take much preparation. But this will not be a hard problem and will be solved easily.

Frank Lunn came down yesterday. He told me he cannot leave his radio business, which he has just started in order to get funds to go to school. I shall ask Bob Johnson to go and write soon as to his answer.

Now about the rings. If you want the double ring ceremony I will get them or rather another to match if possible and have our initials put in them as well as the verse you mentioned. Is all the verse put in or only part of it. Please write as soon as possible concerning these questions. . . . [As] you come to decisions which you have to make, go ahead with the assurance that any decisions you make will have my approval. When we are so far apart, it must be so. . . .

As to keeping it a secret, I no longer feel it necessary or proper, for as God directs us now we will go ahead with our plans as rapidly as possible. I thank Him for you and your friendship. May He bring us together as soon as he wills it possible.

Greet your mother and all your dear ones, for me.

August 29 [AUGUST 28, 3:00 P.M.]

My dearest Walter,

"Now unto the King eternal, immortal, invisible, the only God, be honor and glory for ever and ever. Amen." I Tim. 1:17.

I hope someday we can memorize this verse which I copied from the Bible to give you. To think that our father which is in heaven is the only God and King! May He ever be our Father to whom we will come to at all times.

Walter, counting this day there are but twenty-one days until we are to be married. It seems to me that it would be best to have the 18th to be the date of our day. I can hardly wait and I am sure that it will work out fine. . . .

Last Saturday Beatrice Strand gave me a shower. Two years ago I sang the solos "Because" & "O Happy Home" at her wedding. At the shower she read mother-goose rhymes from which I should be able to get a clue where each gift was hid. I received six bath towels, ~~two three~~ six cannon dish towels, four face cloths, seven yards of toweling for hand towels, a red and white luncheon cloth, and a wear-ever cake pan. Won't these come in handy this fall? I was surely surprised about getting the shower. Ebba and I went to mail some letters and on our way back we said "Hello" to Beatrice. She invited us to stop in for a minute and just as I stepped in the ladies said, "Surprise!" Ebba and mother knew about the affair. . . .

    I am sorry that your knee is swollen. We certainly do not realize how quickly misfortune may come our way. I pray that God will heal your knee. It could have been caused by crawling on the floor since you aren't used to such exercise. I wish that I could have helped you with the cleaning. . . .

    May God bless and keep you.

    Lovingly in Christ, Margaret

    P.S. . . . It will be impossible for me to go to Minneapolis this week . . . the following week Ruth and I are planning to go. I have seen a few pictures of girls who were married in street dresses and I don't like them. I just believe I shall get a white long dress instead. Ruth claims she can buy a dress on sale and then we can change the dress so that it will look nice. She is very gifted in designing dresses. Walter, I can hardly wait to see you. May "He who guideth every star" be our guide always.

August 30 [AUGUST 31, 3:00 P.M.]

My dear Walter,

"Hereby we know that we love the children of God, when we love God and do His commandments. For this is the love of god, that we keep His commandments: and His commandments are not Grievous." I John 5:2,3. . . .

Yesterday I baked four loaves of white bread. They turned out so well that I can hardly realize that I made them. I hope that is the kind of bread-baking that I can do for you. . . .

There are only 18 days left before the wedding. It seems not much has been done in preparation except that the usual house work has been done each day so in that way I am preparing for the days after the wedding. . . .

Yours in Him, Margaret

I received a letter from Helen Foster. She wrote that all except me are returning to teach at Washburn.

Mother heard from Aunt Mary that the man of whom I spoke to you still wants to pay cash for her father's farm. I think the transaction will take place before the wedding. It will be fun to have more of our debts erased.

Walter, I surely miss you. Mother sends her greetings to you.

Dad's and Mom's financial debts were erased when they finally sold their home in Greeley to come and live with my sister Ruth and her husband Ed in Denver. Yet they were never truly out of debt. When they eventually needed full-time nursing care, what money they had left after selling their home was turned over to the state of Colorado to partially cover it. When I think about it, maybe the state owed them for all they had done. So much depends on one's perspective.

Aug. 31 [AUGUST 31, 4:30 P.M.]

My Dearest Margaret:

Oh come, let us worship and bow down,

Let us kneel before Jehovah our maker:

For he is our God,

And we are the people of his pasture, and the sheep of his hand. Psalm 95:6,7

Indeed it should be and is a call to worship and praise, as I awake to behold the power and majesty of god in the huge rolling clouds, the powerful sunlight, the beautiful flowers and endless rolling fields which are so green with growing crops. All these a gift and our privilege to have and use, from our good and gracious Father in Heaven. God sent a beautiful rain yesterday which is very good for the potatoes. I am fortunate in having very good potatoes and this rain helped them much.

Everything is moving forward nicely and I am beginning to look forward to my journey to Minn. which I hope to start on soon. I plan to leave on the Thurs before the 18th or if the work goes well on Wed. eve. If the date is to be the 25th I shall leave a week or less later.

I ordered the rugs or linoleums for the kitchen and will lay them this week. I plan to get a few old chairs from home and then use the little table I have (it is Browns) until we get our own. I will put the bed into the room we have planned for our bedroom and also the old chiffonier [chest of drawers] . . . We can discuss this more when I see you and as we are on our way home.

It is fun to write on practical things, but they will all come out all right and now I cannot help expressing a desire to see you and pray with you and hold you and talk to you. I have so much to be thankful for and I love you, so much.

As I close now, I know that Our Father, holds you and carefully guards you, guides you and keeps you; for He is indeed, kind and gracious, above my understanding and worth.

In Christ, Walter

Now to my Darling, I say; Be always of good cheer, for we can do all things in him that strengthens us. Never be blue (at least no longer than it takes to come to him in prayer) but rejoice always in Christ Jesus.

September 2 [SEPTEMBER 2, 3:00 P.M.]

My dearest Walter,

"Oh magnify Jehovah with me,

And let us exalt His name together." Ps 34:3.

I like the manner in which David invites Christians to praise God together. That's what we want to do during our living together....

I received your letter this morning that mentions your plans of starting on Thursday for Minnesota and of getting some chairs from your folks. Walter, I am so glad that your folks live so near us. I know that your mother will help me a lot....

Frank Lunn can't be along at our wedding. Bernard wrote that he can't; this morning we received a letter from Evodia saying that they can't either. What will we do about all these disappointments. I haven't made any definite plans regarding the wedding yet. Next week I plan to work them out and then I can let you know....

In regard to the verse to be engraved in "our" wedding rings, it is only the first part that I had reference to. I hope my ring will fit my finger. . . . didn't we try the ring on while I was in Colorado? . . .

Last night I felt so disgusted with myself. You see <u>I looked inward and I saw only failures and short-comings </u>but as soon as I emptied my heart of sin on Christ and looked up to Him I found joy, love, light, perfect peace and I knew that all my hopes would someday be fulfilled. This morning I have had a very happy morning and then I received your letter telling me not to feel blue "at least no longer than it takes to come to him in prayer." I surely need you Walter because you direct me to Christ. That's the only kind of husband I want.

The star mail carrier just stopped in and gave us a large bundle of pie plant and eight ears of corn. God does supply us. . . .

Yours in Christ, Margaret

Sept. 2 [SEPTEMBER 2, 10:30 A.M.]

My Dearest Margaret;

. . . I am glad that you decided to keep the 18th of Sept as the date for I am sure that will be much better for us. Under present plans I will arrive sometime Friday afternoon. If there are many things to be done, and I should come sooner, do not hesitate to tell me so. Are there any places near where I could leave my suit to be pressed for it will be hard to keep it in the right press on the way. With much work and packing to do, I believe I better arrive Thurs. evening. Please write and tell me which you think best.

We had a beautiful rain yesterday and last night. It delays the bean threshing but it is certainly a boon to the potatoes. They surely look fine and I believe there is good prospect to sell them at a fair price. . . .

Today we will lay down the kitchen linoleum and put up the stove and move the bed and dresser into the bedroom. . . . I shall have the kitchen and be[d]room furnished (except the curtains) in a livable way. . . .

Well next Sunday and the one following will be the last I will walk through alone (except neither of us are ever alone)[.] The time draws nigh and I grow happier at every thot of it. I do not understand it, I only know I love you until it almost hurts when you are not near. Even my work does not keep my thoughts occupied now. Especially when working around the house. I keep thinking how much fun it will be to plan and work together and share our little worries together and best of all, take them to God together. . . .

In Christ. Walter.

P.S. I went to see Robert Johnson last nite about being best man since Frank Lunn cannot go, but he was not home. I will see him soon and write you about it.

Robert Johnson (no relation to Dad) and his wife were my godparents. They moved back to Iowa shortly after I was born. My parents may have stayed in contact with them, at least through the exchange of Christmas cards, but I don't recall ever meeting them.

September 4 [SEPTEMBER 6, 3:00 P.M.]

Dearest Walter,

"Let us therefore draw near with boldness unto the throne of grace, that we may receive mercy, and may find grace to help us in time of need." Hebrews 4:16.

Isn't this verse a real call to prayer? This morning I didn't get to observe our prayer period in our usual way. We just finished breakfast just in time for it but then it was time for our family devotions and I didn't think it right to be apart from them so instead you and mother and I had our devotional period together. . . .

Tonight I am to sing for the wedding which I have already written about. This week I will have to make definite plans regarding ours. I wish that you could speak with me about the wonderful occasions.

Last Saturday the ladies of the congregation gave me a surprise shower. We received many beautiful gifts but all of the many are useful. There were Pyrex dishes, bed linen, luncheon cloths, a set of dishes & glass to match (to serve 4 people), . . . We truly have much to thank God for.

We didn't stay in Eagle Bend because Mother knew this shower was coming. She kept it a real secret. She sent me down to church to get our dishpan that the ladies had used during the anniversary. I was so surprised to see ladies gathered in the church parlors waiting for me.

Only thirteen more days until the wedding! I presume you will reach here sometime on Saturday. Whenever you come; you are welcome. I can hardly realize it all, but I know that it is all true. . . .

In our blest Saviors love, Margaret

Monday

"Whosoever shall call upon the name of the Lord shall be saved." Romans 10:13

Walter, I am so happy because it seems our wedding is almost planned. Yesterday I saw three wedding ceremonies. The weddings were just beautiful. Each of the brides wore long white dresses. They did look beautiful in them. When John came home he said, "Margaret just has to be married in a long white dress."

Ruth came home this noon. She had bought a white dress that she thought I could use but when we noticed the figures on it not any of us thought it suitable so instead she is going to wear it and mother is going to sew one after the same pattern. Mother is going to cut the pattern this afternoon. She plans to sew it tomorrow and Wednesday. The little flower girls, Rhoda & Katherine, are to be dressed in voile with the dresses made after the same pattern as Ruth's and my dress, perhaps.

Walter, could it be possible for Bernice to be a bridesmaid at the wedding. I think it would be fine to have one of your folks in the bridal party. . . . I am to wear white, Ruth a very pale pink, Rhoda a light pink, Katherine a pale green and so Bernice could choose whichever color she desires that would harmonize. I think orchid would be pretty. I hope to write her about this tomorrow.

We are planning to ask the ladies aid if they will serve ice cream, cake, and coffee to the group. . . .

When can your folks come? We would like to know just about when to expect them. I surely am glad that some can come. Evodia says that they don't really expect to come. I was wondering if some way could be arranged so that one of them could ride with you and maybe one with your folks. It wouldn't cost Reuben so much to ride on the train but it would be nice if they could both ride one way. I know that one of the reasons they are hesitant about coming is because of finances. It would be nice to have Reuben sing "Beloved" and to help John in the ceremony. Philip is to give a sermonette, John to play a flute solo *"Til Friden's Hem"* ["To Home in Heaven"] and be one of the ones who are to officiate at the ceremony, Ruth to be maid of honor. I plan to ask Evodia to play the piano and Ebba, the organ for the marches, Ebba to accompany John and Evodia, Reuben in their solos and Bernice [to] be bridesmaid.

This wedding will be much different than I had planned but we have prayed about it and thought and planned, so I believe it is best to have it this way. . . .

Yours in Christ, Margaret

I miss you very much.

Sept. 7 [SEPTEMBER 7, 4:30 P.M.]

My Dearest Margaret:

But now abideth faith, hope, love, these three; and the greatest of these is love. I Cor. 13:13.

Good morning Margaret, may God's rich blessing be upon you at all times and may he make our love, to grow more and more, for each other, for our fellowmen and for Christ: Even as he has nurtured it thus far, for I love you to the point that it almost hurts especially when you are so far away.

. . .

There need not be a great deal of preparation and I shall probably be up there for Fri & Sat. to enable us to do some of the necessary things.

There is much to do here. The weather has cleared and is very warm (mildly so) and beautiful. My hay is cut and I will probably thresh beans to-morrow and stack hay next Monday. All is going very well. . . .

My darling; I long so, to see you, to speak to you and to hold [you] close. I did not know that love was that way. I miss my folks when I am away, but it seems that a part of me is gone when you are away. But it will only be a few days until we will be together. Let us not forget to pray much about our future. Our planning, our house and God willing our children and our mating, especially <u>the first night. You expressed a concern about it</u>, when you were here and I too am concerned and rightfully we should, tho I am not worried and we need not be if we ask Gods help, which we certainly shall. We shall spend much of that first night in prayer together. In a fair effort to learn as much as possible; to know all I should, I went to Reuben and asked his advice and counsel, and he said that we already had the greatest requirement, Christian love. I thot that the most important all along, so I did not try very hard to read much of what can be read today. Much of it is not reliable anyway and does not consider it from a Christian standpoint. So much for this. It need not and will not cause us any concern, for again I ask you to pray much for us, for yourself and for me.

We are all well and greet you. The folks are planning to come. Bernice & Irene will also go. Greet your Mother, Brothers & Sisters. Now may God always keep you in Christian love in Christ Jesus.

In Christ, Walter

September 8 [SEPTEMBER 8, 3:00 P.M.]

My dearest Walter,

"Thou wilt keep him in perfect peace, whose mind is stayed on Thee, because he trusteth in Thee." Isaiah 26: 3.

This morning we read the twenty-sixth chapter for devotions. The above verse is the one that was outstanding to us. Mother and I both marked it in our Bibles. It is a wonderful verse with a great promise.

I am very glad to hear that Robert can come. I know that you think a lot of him. I have written a letter to Bernice asking her to be our bride's maid. I am so sorry that I didn't think of it before. I hope that she can be along in the procession. . . .

Walter, I think that if you can come Friday afternoon we can easily get our work and packing done. The folks won't hear of our leaving here until after Monday.

Then besides we want to have our wedding picture taken. There is an exceptionally good photographer in Wadena. We can speak of this when you come. . . .

Have you made out your list for announcements of our wedding? I plan to start mine today. Then on Friday, Saturday, Monday, or Tuesday I plan to see the editor to print the announcement. I wrote down four days because I don't know when I shall receive your list. When you are here we can address the envelopes and have them all stamped ready to be sent on Monday morning. Won't it be fun to mail these? . . .

Ruth and Ebba are in Minneapolis picking out a wedding dress for me. Whatever they pick out will fit me fine and will look better on me than if I spent days in Minneapolis looking for one. I can pick out everyday dresses easily but something nice I just can't do well. . . .

Now may God bless and keep you and yours.

In Christ, Margaret

Here are the envelopes (with the same return address) each will receive before they meet at the marriage altar. Examples of the artistic richness and caring beauty conveyed in their handwriting follow along with my transcriptions of the enclosed letters.

Sept. 10th 1938.

My Dearest Margaret:

And the God of all grace, who called you unto his eternal glory in Christ, after that ye have suffered a little while, shall himself perfect, establish, strengthen you.
I Peter 5:10.

The time when I shall see and speak to you again is approaching swiftly. There will be more details tommorrow. I plan to see and speak to Reuben and Evodia. We may come together and I will probably be there no later than Thurs. evening, then. This letter will also come by airmail.

The folks I believe I have told you before, will arrive sometime at.

I had forgotten about making a list of friends to who I would like to send announcements, but this is it. It will be fun to tell the good news. If Reuben goes I must get someone to act as Supt of S.S.

Bernice got her letter and will probably send the answer imediatly.

The wedding plans seem to be coming very well. It could not be otherwise for I know you pray much to a God, who we know has already answered many prayers for us. I cannot on this paper tell you how good our Father has been to me

Clarissa, Minnesota
September 10, 1938

My dearest Walter,

"Fear not, little flock; for it is your Father's good pleasure to give you the kingdom."

How wonderful Christ is! Just think if we had more receptive hearts and minds how much Christ would shower upon us. Tomorrow we plan to go to services here at 9:30; to Eagle Bend services at 11:00; have dinner at Eagle Bend and go to a wedding at 2:00, God willing. Next Sunday, if all goes as planned, you and your folks will be with us. I shall be so very happy to have you all.

Mercy Peterson, Mrs. Ernest Peterson, is to bake our wedding cake. We ate breakfast at her home last Christmas morning. Do you recall? She told me that she wanted us and John's, mother and Ebba for some supper while you are here. She also said she invites you and me to come for breakfast next Christmas. I just wonder where we will be. Evodia and Reuben

Sept. 10 [SEPTEMBER 11, 3:30 P.M.]

My Dearest Margaret:

And the God of all grace, who called you unto his eternal glory in Christ, after that ye have suffered a little while, shall himself perfect, establish, strengthen you. I Peter 5;10.

The time when I shall see and speak to you again is approaching swiftly. . . .

I had forgotten about making a list of friends to whom I would like to send announcements, but this is it. It will be fun to tell the good news. . . .

It is late and I must study my S.S. lesson, take a bath and go to bed. Threshing beans is surely a dirty job. I must [be] up early tomorrow for our prayer hour, the last one apart from one another. . . . It strengthens me every time I come to God in prayer and I know I could not live without it. Therefore I look forward to the day when we will pray together.

Until Morning I bid you goodnight, my darling.

Sunday Morning

Good morning Margaret.

Blessed is the man to whom the Lord will not reckon sin. Rom 4;8.

It is raining gently here this morning and it is very beautiful. I am very contented and very happy this morning, after our prayer hour. I know that you too are. Next Sunday we will have it together. It will be a busy day and we will need God's strength to carry us through it. But he is always able to supply.

There will be another letter as soon as I see Reuben and Evodia today. There is yet much to do before I am ready for Sunday School so I have to close bidding you; God's richest blessing knowing that he always is near unto those who come to him in prayer.

In Christ, Walter

September 10 [SEPTEMBER 12, 3:00 P.M.]

My dearest Walter,

"<u>Fear not</u> little flock; for it is your <u>Father's good pleasure to give you</u> the <u>kingdom</u>" (emphasis in original).

How wonderful Christ is! Just think if we had more receptive hearts and minds how much Christ would shower upon us. Tomorrow we plan to go to services here at 9:30; to Eagle Bend services at 11:00; have dinner at Eagle Bend and go to a wedding at 2:00, God willing. Next Sunday, if all goes as planned, you and your folks will be with us. . . .

Walter, I plan to go to Long Prairie either Monday or Tuesday morning to get our license. I wish that you could "come go along." Mother and I rode to Long Prairie with John this morning. I bought a pair of shoes and a pair of hose for our wedding. Since Ruth could find no dress suitable for me, she bought cloth and now mother has been busy sewing. She hopes to have it finished by Wednesday if not earlier. . . .

As so often happens, mothers are the last resort. At the celebration of Mom and Dad's fiftieth wedding anniversary with the congregation in Greeley, Mom wore the dress being sewn when this letter was written.

> Walter, I am glad that you do miss me. I am glad that we will soon be together. <u>May God bring us</u> more closely in love to one another. I feel that I do love you and I am assured that this love will grow deeper and stronger each day. I long for you and the time when we will be in our own home. <u>May God grant us</u> to earn our daily bread thru the strength and wisdom that He alone can supply. <u>May He grant to us</u> children who may live Christian lives and bring others to His fold. I am so glad that God has revealed to me that you are the one whom God has always intended for me. <u>May we work in His kingdom on earth</u> and do our work because He gives us strength and then when our summons comes [may we] be <u>taken up to be with Him</u> forever.

So many aspects of this "prayer" were answered. God did bring them more closely in love with one another. God did enable them to earn their daily bread. God did grant them children. God did give them strength to work in "His kingdom on earth." As promised in Matthew 6:19, by the time that summons came, heavenly garments had been sewn for them that neither moths could consume nor rust corrode. The spirit of these letters and what I've been able to share of the spirit of their lives are, at best, glimpses of those heavenly garments.

Mary, my sister, was born roughly eleven months after this letter was postmarked; a year and a half after that, Paul came along; another year and a half, James; another year and a month, me; another year and nine months, Lois; another year and two months, Ruth; four years later, Caleb; two and three months later, Anna.

Although each child brings many joys, each creates demands on a parent, especially a mother. Shortly after Ruth was born, Mom was taxed to the point that Lois spent some months at Uncle John's home in Cheyenne, Wyoming. When Lois turned two, she and I spent part of summer and most of fall at Aunt Ruth and Uncle Floyd's wheat farm in Garfield, Kansas. Mom later told me that when I began calling Aunt Ruth "Mama," she and Dad decided to come and get us.

Today I was down to the printing office. On Monday we plan to pick out our announcement. I presume your list of names will be here soon. Mr. Etzel said that the announcements would be ready by Friday. . . .

Monday.

"Count your many blessings

Name them one by one."

Dear Walter,

I have just returned from Long Prairie. Now our license has been bought. It is to come on the Saturday noon mail. I had to swear to the statements I made for you and me. Walter may god bless us during our marriage and always. . . .

Mother is busy sewing my dress. . . . Edna sent me her veil to wear. Ruth has planned how it is to be worn. . . .

Evodia wrote that if they come you will ride in Reuben's car and then after the wedding they are to take the train back to Greeley. I am sure that there is so much for you to do so you can't come on Thursday but whenever you come I'll be waiting for you.

Walter, I am afraid that you don't like the preparations that are being made for the wedding. I am sorry but it seems my folks didn't like to think of us just getting married after services. The people here have done so much for the folks so that is the reason they wanted to have a bigger wedding than just after services. You still love me tho, don't you? . . .

Yours in Christ, Margaret

SOCIETY

Immanuel Church Scene of Wedding

Soft music, glowing candlelight, the hushed quietness of the flower and fern decorated sanctuary of the Immanuel Lutheran church in Clarissa formed the setting for a beautiful wedding ceremony, Sunday evening at 8 o'clock, when in the presence of relatives and friends and members of the Eagle Bend, Clarissa and Little Sauk Swedish Lutheran congregations, Miss Margaret Leaf of Clarissa was married to Mr. Walter Johnson of Eaton, Colorado.

Previous to the ceremony, Rev. John N. Leaf, pastor of Immanuel and brother of the bride, offered a flute solo, "Till Fridens Hem", after which the congregation joined in singing the hymn, "All the Way My Savior Leads Me", and was led in scripture reading and prayer by Mr. John F. Anderson, brother-in-law of the bride. Using the scripture text Eph. 5:22-33, Rev. Philip A. Leaf, brother of the bride, then spoke briefly on the "Christian Home—A Foretaste of Heaven".

Immediately following the vocal solo, "Beloved, It Is Night", by Rev. Reuben E. Carlson, brother-in-law of the bride, the wedding party entered the church and took their places before the altar while Lohengrin's Wedding March was played by Miss Ebba Leaf, sister of the bride. Rev. John N. Leaf, assisted by Rev. Reuben E. Carlson, performed the double ring ceremony and while the couple knelt at the altar Rev. Philip A. Leaf sang the hymn, "O Perfect Love". Following the ceremony the couple marched to the church parlors where they received the well wishes of the large host of friends and where a reception in their honor was held.

Miss Ruth Leaf, sister of the bride; Miss Bernice Johnson, sister of the groom; Mr. Robert Johnson, brother of the groom, and Mr. Gordon Hokanson were the attendants of the bridal pair. The bride was given in marriage by her mother, Mrs. J. P. Leaf. Rhoda, Joanne and Kathryn Marie Leaf served as flower girls, Anna Therese Leaf was veil bearer and Adin Leaf, ring bearer. Mr. Rudolph Aspengren and Mr. Russell Anderson served as ushers.

About 275 guests were gathered for the reception in the church parlors, which had been appropriately decorated with pink and white crepe paper and white bells. The appointments at the tables at which the bridal couple and their immediate relatives and close friends were seated were also in pink and white. The large wedding cake which graced the center of the table had been prepared by Mrs. Ernest Peterson of Eagle Bend. Rev. Daniel T. Martin served as toastmaster of the evening and introduced a varied program of toasts and musical numbers. The arrangements for and the serving at the reception were taken care of by the ladies of the Immanuel congregation.

Mrs. Johnson is the daughter of Mrs. J. P. Leaf of Clarissa and has for the past few years made Clarissa her home. She is a graduate of Bethany college, Lindsborg, Kansas, and has engaged in the teaching profession since her graduation. For two years she was a member of the Clarissa high school faculty. Mr. Johnson is the son of Mr. and Mrs. C. A. Johnson of Eaton, Colorado. He has farming interests at that place and the newlyweds are to reside there.

Out of town guests here for the wedding were: Mr. and Mrs. C. A. Johnson and daughters, Bernice and Irene, of Eaton, Colo.; Robert Johnson, Rev. and Mrs. Reuben E. Carlson and family of Greeley, Colo.; John F. Anderson, Augustana Seminary, Rock Island, Ill.; Miss Ena Mattson of Braham, Miss Selma Huseby of Dilworth, Rev. and Mrs. Philip A. Leaf and family of Mora, Rev. and Mrs. Daniel T. Martin of Minneapolis, Rev. Lambert Engwall of Parkers Prairie and Rev. and Mrs. Gunnar Goranson of Wadena.

A write-up of the wedding in *Lone Prairie Leader*, September 20, 193

Walter and Margaret—my mom and dad—on their wedding day, September 18, 1938. Mom wore this dress at the celebration of their fiftieth wedding anniversary.

A photo taken around 1970 of Mom, who saw God in all things, and Dad, who had faith in God at all times

A Glimpse Back

August 15. "Thou wilt keep him in perfect peace whose mind is stayed on thee" so rang in Margaret's ear that she wanted to share it with Walter. What spiritual anchors ring for you?

August 18. Consider Walter's statement, "If all the way our Savior leads us, we certainly do not have much to ask beside." It is a logical statement with an assumption regarding priorities and a conclusion regarding consequences. Can you imagine circumstances in which it would aid him in making difficult choices? If so, in what ways? What words would you use to summarize the most basic priorities governing choices in your life?

August 19. How realistic is Margaret's feeling they will be living like a king and queen if they get a "good kitchen set," "a good range and a comfortable bed," and "a piano"? What would you need so that you could live like a queen or king? What might make you want more once you got what you needed?

August 25. If "God knows" before the choice is made, how does your sense of personal choice deal with that—a contradiction in terms, a paradox, a conviction that there is a deeply satisfying choice, or an assurance that the eventual choice will be the right choice?

August 28. Walter sees the Sabbath more as a gift from God than as a rule to "pin man down" on a particular day. In what ways can spiritual understandings and injunctions be turned into and viewed as laws and regulations that take the joy out of life as opposed to gifts that bring joy into life?

September 2. Margaret mentions looking inward and seeing "only failures and short-comings," but the next day, on emptying her heart on Christ and looking upward, she "knew that all [her] hopes would someday be fulfilled." Margaret does what Walter suggested in a postscript he wrote the day before the night she felt so disgusted. Is that purely a coincidence, or is it a consequence of an emerging harmony in their spirits?

September 7. Many couples have concerns about their "first night" together. For Walter and Margaret, that night will be the same day they are married. Walter is hoping it starts out with prayer. What do you think should concern a couple on that first night, and how should that concern be addressed?

September 10. In her last letter, Margaret ends the courtship duet that she has been singing with Walter with some requests that

I've underlined. In the discussion of the embodied and expressed spirits at play in these letters, we encountered the guiding spirits of mindfulness, truth, and service. How do you see those being involved in Margaret's obtaining what she wants in her requests? How might they be involved in obtaining what you want in life?

To summarize what these letters have meant to me will take more than a song. As a consequence of the Night of the Little Self and how I came to see the guiding spirits contrasted in our scriptures, a contest between the voice of the competitive big self and the voice of the inclusive little self was needed. After publishing the first volume of this trilogy, I became interested in the altruistic and egoistic voices singing in the field of evolutionary psychology. This additional distinction resulted in a competition in four acts filled with suggestive harmonies and dissonances each time the holistic voice of the little self was joined with the altruistic and egoistic voices of the big self.

The stage for the contest is set in the first act, "What Lives On."

In the second act, "The Universality of Our Most Basic Spiritual Choices," the three voices differentiate themselves by the tunes they promise to sing for common concerns they all share—life, disclosure, worthiness, relationships, brokenness, and death.

These common concerns and the trios that result comprise the different scenes of the third act, "Consequences of the Elemental Guiding Spirits We Heed."

The voices end up disagreeing on humanity's fate in the last act, "Assuring Life in the Hereafter."

Those who can enjoy the contrast of voices in opera should find the next and final chapter, "Joy in This Life and the Life Hereafter," interesting and thought provoking.

CHAPTER 8
JOY IN THIS LIFE AND THE LIFE HEREAFTER

To use Margaret's verb in the last of her courtship letters, Mom was "summoned" by God to join the saints in heaven on April 4, 1999, and Dad on February 1, 2001. Over the years, we children had gone our separate ways, as had many from the congregation of which my parents were so much a part. Consequently, a heavenlike ambiance of spiritual reunion pervaded the congregation's celebration of Mom's and Dad's now full union with God, a congregation Dad had known his entire life and Mom her sixty-one years of married life.

I was asked to lead the remembrance portion of their funeral services. In addition to performing some musical numbers, my brothers and sisters and I introduced our families and shared some thoughts and recollections. With saddened but thankful hearts, we sang "Children of the Heavenly Father" for the congregation, a favorite of Mom's, and "Mine Eyes Unto the Mountains," a favorite of Dad's. Remembrances were shared during the service and for the next hour or two as the congregation lingered over refreshments. All the while, the words of Jesus kept washing over my mind: "Well done, good and faithful servant; you have been faithful over little, I will set you over much."

Well done with respect to what? For the time being, I settled on the ever-present concern of my parents for what made for peace and joy for themselves and for others in their mortal lives and in the life to come. But what was the "much" over which their faithfulness would now be set? That puzzled me for some time. It clearly had to do with the *kingdom* of God (to use a designation of the realm of God found in Luke 13, verses 18 and 19, of the Inclusive Bible, which captures our oneness with the creative spirit and the web of humanity) that Jesus proclaimed and that had attracted their embodied spirits. That was spiritually evident and is readily inferred from these letters. It was materially obscure at the time, for our parents left us very little financially. I could imagine some sort of angelic responsibility they might have in heaven but not an increased responsibility they might have on Earth.

  A few years after their funerals, we drove to the cemetery plot Dad and Mom had purchased in Greeley to bury their ashes. We exchanged some remembrances, shared some prayers, and sang some hymns. Then we moved to a brass marker inscribed with their names and life dates. It lay between two large cottonwood trees, much like the ones we had climbed as kids. We deposited some of their ashes and headed for the Rockies. After stopping at Jax Surplus in Fort Collins, where, as a family, we had often bought fishing tackle, we drove to Ted's Place. We purchased our usual loaf of sandwich bread and a couple of cans of Spam before heading up the Cache la Poudre Canyon.

  A half hour later, we stopped along a broad stretch of the river, piled out of the cars, and climbed down to the river. We skipped some rocks over the water before getting out the Spam and sandwich bread. The family's joking about the additives in processed meat didn't deter my brothers and me from sharing a last fishing lunch with Dad.

The lunch site was cleaned, more hymns were sung, and more remembrances were shared before each of us sprinkled some of their ashes on the river. When my turn came, Chloe, my six-year-old granddaughter, walked out with me to a rock around which the current swirled. When I opened the canister, she peeked inside.

"Are those your mother's bones?"

"Yes, they are bits of her bones. These bits are now free and headed for the ocean, but my mother is not in the bits. She is in heaven as surely as she is here in my heart."

My answer reflected the comforting relationship I had with my parents. Maybe it satisfied Chloe, as she didn't question me when heading back. However, it revived my interest in Jesus saying that those who have been faithful over little would be set over much now that my parents' ashes were floating to the sea. Answers started coming years later as I puzzled over how Walter, who wasn't in Washburn and had never been, had changed the conversations the day Margaret displayed her engagement ring at school. In time, those answers gave rise to the song "Watching the Embodied and Expressed Spirits at Play in These Letters." To see how that song relates to our joy in this life and the life hereafter, we must look more closely at what lives on after we physically are no more.

What Lives On

As noted earlier, your embodied spirit resides in your mind, but your expressed spirit resides in the minds of others. To better understand what aspects of your expressed spirit exist in the minds of others, how extensive they are, and how they exist, we must look more carefully at your expressions and their consequences.

Your expressions convey the understandings of your mind, the feelings of your heart, the health of your body, and the wellness of your soul. These personal truths come across in your shape and walk, the clothes you wear, the surroundings you prefer, the places you frequent, the performances that interest you, and the activities you favor. They come across in those you accept, reject, or ignore; in the groups you join; and in the leaders you support.

Your direct expressions are experienced by those who physically see and hear you. They are felt by those you hug or push; they are inhaled in your presence and savored at your table. Young or old, rich or poor, sick or well, happy or unhappy, the thoughts and choices comprising your truths go forth and, as I experienced on the Night of the Little Self, at prodigious rates.

Your truth expressions have consequences that can change the minds of their recipients. The changes reside in our conscious memories and subliminal associations. Although sophisticated scanning technologies can differentiate neural patterns in our brains, they can't distinguish the distinctive character of your thoughts and visions—at least not yet.

A sense of the significance of your expressions on others can be gained by considering the consequences that their expressions have for you. You may have been lifted up by an appreciative hug or cast down by a cutting criticism. Your hopes for a caring world may have been deepened by a stranger's unexpected kindness. Another's hovel, home, or mansion may have set you thinking about fairness. Someone's teaching may have inspired you. A conversation may have set your life on a new path.

An appreciation of the extended consequences of what you have said and done can be gained by contemplating the sources of the expressions of others that led up to a particularly expressive moment of yours—especially those that have left you pondering the diverse coincidences that brought it about. I shared a few earlier that led up to my spiritual inversion on the Night of the Little Self. Reading the Bible at my mother's urging set me pondering the nature of God; college discussions sharpened my questions; delving into the book *Mind, Self, and Society* after noting it on my brother's bookshelf started me looking at life from a totally new vantage point.

Your expressed spirit incorporates aspects of the expressions of all who touch your life. Although Margaret's letters were meant for Walter, you can't read them without imbibing aspects of the expressed spirits of others that reside in her. When she writes, "Mother is busy sewing my dress," a bit of her mother's serving spirit may bring your mother to mind. When she writes, "Ida just arose and is busy peeling potatoes. She said that I should continue writing," a bit of Ida's spirit may have set you contemplating the source of that helpful spirit, which may lead you to pass Ida's helpful spirit on to others. And on it goes. Although Margaret's embodied spirit principally guides her writing, the influences of the expressed spirits of many are being conveyed in the spiritual seeds she sows in these letters.

The reach of her words and actions comes about via the relational web over which we send and receive the hosts of expressions that enliven us. There are outgoing links to those in our communities who can perceive our direct expressions and incoming links conveying the direct expressions we perceive. There are generational links within families, interactive links among friends, professional links within organizations, directed links from writers to readers and from leaders to followers, command links within armies, and fluctuating links between those on opposing sides of a conflict. Who knows where and how throughout this vast web the spiritual seeds Margaret has sown in these letters and throughout her life will land and take root in others? When taking root, they will grow and bear even more seeds that will, in turn, get sent throughout the web. Who knows to what extent and in what forms this will all take place?

It was the bits of the expressed spirits of my parents residing in others that gave rise to the spirit of their funerals and largely set the topics and tone. The pervading joy in having known my parents came through in gratifying conversations that kept bringing back the words of Jesus: "Well done, good and faithful servant; you have been faithful over little, I will set you over much." The "little" over which they had been faithful came through in the anecdotes. It was when contemplating how the spirit of their expressions had been spreading over the relational web that I began to grasp the "much"—and that was before their courtship letters surfaced.

Your expressed spirit is finally birthed when death cuts its last tether to your mortal mind. Much as a baby must be fed another way once its umbilical cord to its mother's womb is cut, so your expressed spirit must be "fed" another way once its "umbilical" connection to your direct expressions is no more. The only way that can happen is through the direct expressions of others.

Although the achievements of community leaders often fill the news, it's the memorable anecdotes and takes on life that enliven the remembrances of those close to the departed. To understand the significance of these anecdotal remembrances, it helps to distinguish the spirits of our expressions from the details of those expressions. The consequences of an intense conversation between two individuals that "cover the waterfront" of their relationship are illustrative. In such an interaction, countless visual, auditory, and physical expressions are exchanged. Their immediate consequences get sorted out when the participants go their separate ways and reflect on what took place. Because of the profusion of what was said and done, factual details quickly fade and are forgotten (unless somehow recorded); the spirit lingers on, possibly linked to a few memorable details. We later share those spiritually encapsulating details with concerned friends and sympathetic listeners.

We naturally miss the physical presence of the embodied spirit of a loved one at a funeral. However, when their expressed spirits are alive and well within us, we soon find ourselves anecdotally resharing what their lives meant to us. Paula Johnson, my cousin through marriage to Barry, shared one at my mother's funeral when she recalled the birth of their son Aaron. A serious handicap resulted in Aaron's living a short life. That life while lived deepened Paula and Barry's joy in their own lives—but not when Aaron first came home. The sympathetic tears and hugs of their friends seemed to make things worse. Paula said that Aunt Margaret and Uncle Walter didn't offer sympathy when they came to see her. Instead, she shared how Aunt Margaret suggested that they "kneel and pray for God's strength in caring for Aaron." In their praying together, Paula said she felt an unexpected comfort and strength.

Paula's anecdote is noteworthy in the concerns and guiding spirits at play. There is the sadness that Paula must have felt in seeing nothing but difficulty lying ahead in her life with Aaron. Her friends came with a helping spirit evoked by their relational concern for her. Paula found their visits discomfiting though well meaning, possibly in their feeling sorry for her. Walter and Margaret also came with a helping spirit and a relational concern for Paula. Yet she found their visit strengthening and comforting, possibly in the turning of her attention from a self-pitying spirit to a guiding spirit that could help her.

Differences such as these in the consequences of the spirits we heed will become clearer in the upcoming verbal contest between our egoistic, altruistic, and holistic self-orientations, but first we must develop a basic and commonsense understanding of the basic concerns giving rise to our guiding spirits and the self-orientations to which those spirits variously appeal. This will be done in the next act, which brings together a religious and a scientific perspective.

Our Most Basic Spiritual Choices

The innate and distinctly human character of the moral issues in our basic spiritual choices has been pointed out in both our religious and scientific institutions. In the Abrahamic religious tradition, humankind was set apart when God said,

> Let us make humankind in our image, according to our likeness; and let them have dominion over the fish of the sea, and over the birds of the air, and over the cattle, and over all the wild animals of the earth, and over every creeping thing that creeps upon the earth.
>
> Gen. 1:26, New Revised Standard Version (NRSV)

God uniquely privileged humankind when He enjoined, in Genesis 2:16–17, "You may freely eat of every tree of the garden; but of the tree of the knowledge of good and evil you shall not eat, for in the day that you eat of it you shall die." After eating of that tree, we were told in Deuteronomy 30:14, "the word [the knowledge of good and evil] is very near you; it is in your mouth and in your heart, so that you can do it."

In the theory of evolution, the distinctiveness of humankind is also recognized. In *The Descent of Man*, Charles Darwin wrote,

> The differences between man and the lower animals, the moral sense is by far the most important. . . . It is summed up in that short but imperious word *ought*, so full of high significance. It is the most noble of all the attributes of man. . . . The moral sense follows, firstly, from the enduring and ever-present nature of the social instincts, secondly, from man's appreciation of the approbation and disapprobation of his fellows; and thirdly, from the high activity of his mental faculties.
>
> *Great Treasury of Western Thought* (*GTWT*), 582
> (emphasis in original)

Both narratives deal with issues that have been around a long time. Both have notions of "good" and "bad." The biblical account suggests a good within our heart that identifies with the creative spirit of humankind and with all over which we have dominion. Let's term that sense of self the *holistic self*. It is what I experienced and called the little self the night I was caught up in the unfathomable relational web of the creative spirit.

The Darwinian account suggests a moral sense of good "summed up in that short but imperious word *ought*." In his book *Evolutionary Psychology: The New Science of the Mind*, 6th ed., David Buss discusses the advances being made in understanding how this "ought" sense of self could have evolved. Those advances resulted in a broader notion of evolutionary fitness that genetically rewarded selfless behaviors, such as that of a young man willing to risk his life for his country. Let's term that resulting sense of self the *altruistic self*.

The third is a sense of self that both accounts deprecated, one with "evil" and the other with "lower animals." There is an uncaring sense of self-centeredness in both the notion of "evil" in the biblical account and in the survival-of-the-fittest notion of the Darwinian account. Let's term this sense of self the *egoistic self*.

Although all three self-orientations were alluded to in that Night of the Little Self, I lacked the words for what did and didn't change other than an old, competitive big-self view of life being subordinated to a new, inclusive little-self view of life. Words for the change in the spirits I would be heeding came through when I read through the Bible again. Words for what didn't change came when I realized, for example, that a person heeding a truthful spirit and a person heeding a deceitful spirit had a common concern: the nature of their disclosures. Neither did my basic concerns with life itself, with worthiness, with what to do when things went wrong, and with what of me lived on after death. We all have these universal concerns, regardless of the spirits we heed when addressing them.

There are many words for sharing these concerns and the spirits they evoke. They enrich our stories and enliven our songs. The names for these concerns and related spirits are filled with nuances and shaded with connotations. Because these spirits and concerns are relationally interdependent, assigning them common yet distinguishing names is fraught with ambiguity. Those that worked best for me when composing the songs to be sung in Act 4 by our self-orientations are given in Table 1.

Table 1: Our Elemental Spirits Organized by Universal Concern and Self-Orientation

Concern	Egoistic Self	Altruistic Self	Holistic Self
Life	Self-aggrandizing	Promoting	Mindful
Disclosure	Manipulating	Consenting	Truthful
Worthiness	Ranking	Grouping	Encompassing
Relationships	Dominating	Role playing	Serving
Brokenness	Faultfinding	Help seeking	Self-searching
Death	Dying	Leaving	Birthing

We encounter these elemental spirits each time these universal concerns are instantiated in the issues of our lives. The ones we heed are central to our finding or not finding joy and meaning in life. The reasons will become apparent in the next act, which lays out some of the consequences and moral implications of the spirits we choose to heed.

Although the details of our personal concerns will reflect the circumstances of our lives, the spirits we heed and the underlying concerns in Table 1 do not. They have survived the ages and no longer depend on time.

When overwhelmed by the pressing matters of the day, we seldom pause to think about the consequences of the spirits we have been heeding and the basic concerns that gave rise to them. We should. It is easy to overlook opportunities that have been missed for finding more joy and meaning in life when we don't take the time to pause and reflect.

In their weekly desire to truthfully share their lives with each other, Margaret and Walter clearly have taken the time to more fully understand the personal concerns they are addressing. Their letters provide a facile and reasonably objective means by which we can see the diversity of contexts in which the concerns in Table 1 entered their daily lives while courting. Appendix 1 names some of the personal concerns addressed in their letters. Those chosen were the ones noted in the Glimpse Back sections in each of the books of this trilogy.

Though the naming and classifications of these personal concerns are highly subjective, a couple of things stand out. First is the relative frequency in which the six universal concerns are represented: disclosure (38), brokenness (29), life (26), relationships (20), worthiness (10), death (2). This is to be expected. Disclosure came first, probably from Walter's and Margaret's wanting to get to know each other. That brokenness came next may seem surprising until we recognize how faultfinding, conflict, and tragedy dominate our news channels. That worthiness is so low on the list may reflect what I came to see as their broadly based joy in life. That death comes last as an ultimate concern probably reflects its seldom being an ultimate concern when courting.

Second, naming a personal concern being addressed by an individual helps us decide if we might be interested in the issues but may not specify the underlying concern. If we look at the twenty personal concerns involving their relationship, disclosure was the underlying concern eight times, relationships seven times, and brokenness nine times.

Circumstances are complex. Our sense of what captures our attention at any particular moment can be as fleeting as our minds. Yet it is these matters that we share with one another. From them we can often infer the underlying spirit of a thought or action.

The expressions in Appendix 2 correspond to the personal concerns tallied in Appendix 1 for this third volume. If one looks at the four entries labeled "Experiencing God" in the first and second pages of Appendix 2, one can easily infer the varied contexts under which God is experienced, two in which life was the ultimate concern, and two in which disclosure was the ultimate concern.

Per Table 1, each universal concern can give rise to three different spirits. A personal concern does not specify which of the three spirits to heed. Which one we heed will depend on our desires and the personal goals that govern our self-orientations. Although priorities are seldom stated, even as children, we can often grasp the spirit that a person is heeding from their expressions, as suggested by Table 1.

Walter and Margaret were predominantly addressing their personal concerns from a holistic orientation when writing these letters. That is easily seen from the expressions in Appendix 2. They are ordered first by the relevant row in Table 1 designating a universal concern. Those expressions relevant to the same universal concern are ordered lexicographically by the personal concern being addressed. Those relevant to the same universal concern and same personal concern are ordered by the date (month/day) of the letter with an underline signaling the discussion of that expression in the following Glimpse Back section. Thus, the third row of Appendix 2 contains an expression Walter wrote on September 7 addressing a personal concern (Enjoying marriage) in which Life (the first row in Table 1) is the underlying concern.

It is easy to see that all of the ten expressions in Appendix 2 that entail life as the underlying concern could have been written by holistically oriented persons heeding their mindful spirits. None are reasonably seen as coming from someone heeding a self-aggrandizing spirit. Expressions 7/5 and 9/10 could possibly be seen as coming from anyone interested in the kingdom of God when heeding a promoting spirit. The remaining seven expressions reveal no clear promotional intent beyond their own interests.

Looking at their expressions when disclosure is the motivating concern, they are simply heeding their truthful spirit by trying to accurately convey the thoughts on their mind. We do not see the weighing of pros and cons and the exchanging of approvals and disapprovals that typify the consenting spirit. However, the fourteen expressions by themselves do not rule out their heeding a manipulating spirit of disclosure. That spirit can and will say anything that serves its egoistic purpose. It can only be ruled out when we know that the person communicating the thoughts was not egoistically oriented. Even had we not read Walter's and Margaret's letters, that confidence would have grown with our ruling out the self-aggrandizing spirit in the preceding discussion in which life was the ultimate concern.

This appendix has only two expressions in which worthiness is the motivating concern. Both involve acquaintances whom they perceive as having difficulties brought about by choices related to social temptations. Had Margaret and Walter been heeding their ranking spirits, they might have called them losers. Had they been heeding their grouping spirits, they might have dismissed them as morally, socially, or spiritually unworthy of their interest. Instead, an encompassing spirit comes through in their expressions of concern.

When relationships are our motivating concern, we can heed a serving, role-playing, or dominating spirit. There is no trace of the last spirit in the seven tabled expressions of Walter and Margaret. Their awareness of our having and playing different roles in life comes through in their first (7/12), middle (6/28), and last (5/17) expressions. Few question the need for and benefits of heeding our role-playing spirits in the game of life. It is when we rigidly heed our role-playing spirits and fail to appreciate the role-playing spirits of others that dissensions arise. That possibility is lessened when we heed the serving spirit, as Margaret and Walter do in their remaining four expressions.

We naturally seek positive involvement when the ultimate concerns are life, disclosure, worthiness, and relationships. Involvement in which the ultimate concern is brokenness comes to us on its own. Margaret did not seek to be bedridden in her 7/23 expression. Walter did not relish how his country's privilege of religious freedom was being celebrated in his 7/4 expression. Margaret's feeling disgusted with herself wasn't the result of a conscious initiative in her 9/2 expression. Our joy in life is dampened when brokenness comes our way. Our faultfinding spirits are helpful in pointing out an immediate cause: Margaret's having to lie in bed, Walter's not liking how the privilege of religious freedom was being celebrated, Margaret's feeling disgusted with herself. In all three situations, their self-searching spirits turned their eyes toward something that brought back their joy. Although it seemed to heal Margaret's disgust with herself, it didn't immediately change her having to stay in bed. It didn't change how the privilege of religious freedom was being celebrated in America, but it did release a thankfulness in Walter's own heart for that privilege.

Death becomes a motivating concern when we contemplate what part of our expressed spirits lives on after we die. Each of the three elemental spirits are judgmental in their guidance. The judgments will largely reflect the guiding spirits we have chosen to heed. Margaret expressed her uncertainty in that judgment for her student by writing "I hope he was saved." The song she chose to sing reflects her belief that the judgment lies in the expressed spirit birthed by a "small child of Bethlehem, the unknown young man of Nazareth, the rejected preacher, the naked man on the cross, who asks for [our] full attention" (*You Are the Beloved: Daily Meditations for Spiritual Living* by Henri Nouwen, compiled and edited by Gabrielle Earnshaw, 383).

My parents' letters are filled with their understandings of the expressed spirit of that Nazarene. His teachings came to me in a new light following that revelatory Night of the Little Self. A selection of the writings of his followers related to the distinctions made in Table 1 are given in Appendix 3. They should resonate more clearly with what is being sung in Act 3 once that relevance is clarified.

Our concern for life and the creative spirit in which it is sourced comes through in the first six passages in Appendix 3. We are being taught by example to heed the mindful spirit of Jesus in the temptation context of passage 1 and are taught not to heed the self-aggrandizing spirit of Herod in the judgmental and mocking context of passage 6. Passages 2 and 3 convey the importance of a mindful awareness of our personal connection with the creative spirit of life. That the mindful spirit brings rest and eases life is explicit in passage 3. That it brings trust and joy in life is implicit in passage 4 (a joy expressed on the night of the Last Supper with his disciples when in John 15:11 he tells them, "These things I have spoken to you, that my joy may be in you, and that your joy may be full"). In passage 5 we are commanded to be mindful of the Creator of the universe. Both passages 4 and 5 convey the importance of the encompassing spirit of worthiness soon to be noted.

The next six passages on disclosure assume we know the difference between telling the truth and telling a lie. Passages 1 and 2 stress the importance of abiding in the spirit of truth. Passage 3 emphasizes the importance that the spirit of truth places on whatever needs to be disclosed. Passage 4 conveys what Jesus felt needed to be disclosed about his own life. Passage 5 recognizes the difficulties that can arise when someone heeding their truthful spirit shares expressions that run counter to those of a group heeding their consenting spirit of disclosure. (Later, Jesus stressed the importance of his spirit of truth, saying, "For whoever would save his life will lose it, and whoever loses his life for my sake will find it.") We are warned in passage 6 of the tragedies that can result when a group's disclosure spirits are forcibly captivated by leaders heeding their self-aggrandizing and manipulating spirits of disclosure.

In the first passage of the three on worthiness, Jesus clearly elevates the self-worthiness that comes with the encompassing spirit as a consequence of our being children of God over the transient worthiness that comes with the grouping spirit. Implicit in that passage is freedom from our pursuit of the rewards of the groups with which we identify. The ranking spirit is disparaged, and the encompassing spirit is emphasized as a means of understanding true worthiness in the second passage. (It is unclear what might be meant by "in my name" if not "because of a child's inherent worthiness.") In passage 3, Jesus inverts the ordering of worthiness as seen by our ranking spirits.

This last passage also contains the spirit of service Jesus emphasizes in the next four passages regarding our concern for relationships. The parable of the Good Samaritan in the first one begins with the fear and hurt that the dominating spirits of robbers can engender. Then it points out how our group-worthy and role-playing spirits can keep us, and especially our leaders, from taking the time to address what is individually needed in our relationships. Finally, the encompassing spirit of a man draws the Samaritan into a position where his serving spirit can personally help the injured man. The enabling nature of that encompassing spirit when it comes to service is emphasized in passage 2. The importance of the spirit in which service is rendered is stressed in passage 3. In the fourth passage, we are shown how our role-playing spirits can blind us to what is of lasting value when money and success become our foremost desires.

The next four passages address our spirits for assuaging brokenness in our lives. Without going into detail, passage 1 addresses the problem of a person with a help-seeking spirit repeatedly asking for forgiveness. Our being urged to forgive someone over and over implicitly conveys God's ever-present willingness to grant someone's true desire for reconciliation. The next passage conveys the role of sincerity in conveying that desire for reconciliation and the role of faith in our seeing and feeling the ways in which it comes about. In passage 3, Jesus emphasizes how the faultfinding spirit can direct us away from the very part of the problem that our self-searching spirit can assuredly address. In passage 4, Jesus compares the brokenness that can result from two persons who have too long heeded their role-playing spirits, a tax collector and a religious leader. He first points out how the faultfinding spirit prevented a religious leader from seeing how his group-ranking and self-aggrandizing spirits were keeping him from experiencing the hallmark of his profession: finding oneness with God. Jesus then points out how a humble and self-searching spirit can bring back the joy of reconciliation with God, even for a tax collector.

The three passages related to death, the last universal concern, bear on what is lasting in our expressed spirits. In the first passage, the first two "servants" are rewarded for making full use of their given talents, presumably by heeding their mindful and promoting spirits and their role-playing and serving spirits. The last "servant" is cast away for not using—even hiding—his given talent, presumably because of his egoistically selfish slothfulness. In the parable of the banquet of the second passage, those who felt that what they gained by heeding their role-playing spirits was more important than heeding the holistic spirits of the kingdom that Jesus proclaimed were disinvited. Instead, the less successful, "the poor and maimed and blind and lame," and those in the "highways and hedges" were invited so that the master's house might be filled for the wedding celebration. In Matthew's 22:2–14 version of this parable, one of the attendees is bound and cast into the outer darkness for not wearing a wedding garment to the banquet.

Paul stressed the lasting nature of what we are to wear when he wrote, "For this perishable nature must put on the imperishable, and this mortal nature must put on immortality" (1 Cor. 15:53). Just as there are many ways we might dress when attending a formal affair, the beatitude of Jesus in the third passage suggests the importance of our guiding spirits when it comes to dressing up for this eternal celebration to which all have been invited.

We now turn to how these universal concerns, guiding spirits, and self-orientations play out in our lives. Although some of what follows may challenge what you have been taught, all should resonate with your common sense.

Consequences of the Elemental Guiding Spirits We Heed
Life: Our Concern for Continued Being
When it comes to the spirits arising out of our concern for life itself, contemplating the universe and the creative spirit through which it came about is a good place to start. That universe encompasses more than a trillion galaxies. Our galaxy, the Milky Way, is but one—a single galaxy with hundreds of billions of stars. Our star, though average in size, exceeds our Earth a million-and-a-half-fold in volume. Although negligibly small in that light, our Earth brings about the seasons of our years in its orbiting and the rising and the setting of our days in its axial spinning. Upon its outer film thrive the plant and animal kingdoms of the Earth, of which humans are really only a small part—billions of us interactively linked in a relational web over which our expressions are shared.

But what have these humbling scales and numbers to do with daily life? Could they be nothing more than simple facts to the tens of billions of interlinked neurons making up our brains and signaling each other at numbing rates? Need they be humbling once that neural web realizes it heads a complex body of cells a hundred times its size? Possibly not, once it realizes that body is at its beck and call, a body it can tell to diet or indulge, to exercise or recline, to work or loaf, to attend to or ignore. And should it find within itself an ability to sway the wills of crowds, what then? Such questions are helpful when contemplating the spirits arising out of the most basic of our six universal concerns: life.

We are inherently social creatures, born with markedly different minds, talents, and constitutions. We are reared in vastly different homes and circumstances through no choice of our own. Each of us has been given a latent instinct to excel, possibly for no other reason than our survival. Life's complexity, our diversity, and the relativity of our judgments assure the existence of situations in which each of us will excel and those in which each of us won't, and to markedly different extents.

To comfortably grasp what is going on in all this complexity, it is easy to start off seeing life as an open-ended game in which there are winners and losers. The rules are many, complex, and variously understood. When looking at life through either the altruistic or the egoistic eyes of the competitive big self, possessions, influence, and prestige provide appealing and quantifiable criteria by which to discriminate winners and losers. Life for the big self is a long series of up-the-hill and down-the-hill roller-coaster rides of successes and setbacks, each with its attendant and transient feelings of happiness and sadness. Dean Kay and Kelly Gordon put a captivating spin on that ride in their song "That's Life," popularized by Frank Sinatra (https://www.youtube.com/watch?v=TnlPtaPxXfc).

Competition is good, but when the self-aggrandizing spirit takes over the play of the game, it morphs into a struggle for personal superiority. When winning becomes foremost in our minds, the self-aggrandizing spirit seeks to make its will the law—the "law of the jungle." We become greedy; fair play no longer makes sense. Its modern economic and political manifestations were highlighted in Baruch Spinoza's *Ethics* over three centuries ago:

> In the natural state there is nothing which by universal consent is good or evil, since everyone in a natural state consults only his own profit; deciding according to his own way of thinking what is good and what is evil with reference only to his own profit, and is not bound by any law to obey anyone but himself.
>
> *GTWT*, 690

When heeding our self-aggrandizing spirits, material presence is our strength—its loss our downfall. That material presence can take many forms: physique, dress, wealth, reputation, position, and associates, to name a few. When winning and feeling superior, we are happy and can mistakenly see our talents, efforts, and accomplishments as the determining factors. Should the world not go our way, we complain and blame others and outside factors for the irritations that set in. And they will. The overt expression of our overriding desire for personal superiority can irritate even friends and colleagues, especially when we are purposefully intimidating.

When we get caught up in vocational and recreational activities that require cooperation in our regions of the web of humanity, we are naturally drawn to the promoting spirit of the altruistic orientation. Both the competitive and cooperative tendencies of that self-orientation are emphasized in the team behaviors that should, and usually do, typify athletics, economics, and politics. The big self's innate desire to excel is encouraged when cheered on by "the team." Being a team member can bring feelings of peace, security, and meaningful involvement, whether in a family, a school, a place of employment, a political party, or a nation.

Competition for positions within groups appropriately flourishes when circumscribed by unifying notions of membership, complementarity, and team spirit. Exciting and self-motivating competitions between groups naturally result when kept within bounds by watchfully enforced rules and conventions that are fair and engaging. The far-reaching success of these endeavors has brought about the distinct role we now play and the responsibilities we now have in the plant and animal kingdoms. That success reflects the creativity inherent in the altruistic big self's typifying desire for social influence. When seeking social influence, your political stature is your strength—its loss your nemesis. That stature can also take many forms. Popularity, achievements, wealth, reputation, positions, and supporters are among the more prominent.

Our promoting spirits assure the emergence of trusted leaders, committed participation, fitting rules and conventions, and commensurate rewards. Their impact over evolutionary time is evident in the singular role humankind is coming to play in the plant and animal kingdoms. Yuval Harari alluded to the power of this spirit in his book *Sapiens: A Brief History of Humankind* when writing, "Sapiens can cooperate in extremely flexible ways with countless numbers of strangers. That's why Sapiens rule the world, whereas ants eat our leftovers and chimps are locked up in zoos."

Both the appealing benefits and attendant risks of the promoting spirit's ability to accomplish great things have long been recognized. Both were tellingly captured centuries ago in the story of the building and destruction of the tower of Babel (Gen. 11:1–9) and in the Samuel and Kings narratives of the rise and fall of the Davidic kingdom. Scientific understanding and technological advancements are now markedly increasing both the benefits and the risks. Both dominated the news when Margaret wrote "the clouds of war are lowering over us" on November 11, 1937 (*EG*, 53), a dozen years after Hitler began his ascent to power. How unswervingly supporting self-aggrandizing political leaders can result in the horrific and unthinking birthing of our expressed spirits will become clearer when we discuss our concerns for worthiness and disclosure.

Although the mindful spirit appreciates the many things that we have done and can do, it much more humbly recognizes and appreciates the unbelievable depth and detail of what the creative spirit has done and will continue to do throughout the eons of astronomical time. The mindful spirit seeks to be in harmony with the incomprehensible coordination in which it is caught up; it does not seek to control it. In seeking the reaches of that harmony, it reawakens the quickening curiosity we once had as children. It recognizes the inclusive reality in which we live and how each of us has been variously invited, gifted, and positioned to do our part. In heeding the mindful spirit in our concern for life, all are called to participate in groups working for the common good; all are called to appreciate the roles played in furthering the web of life; all are called to be mindful of their need to efficiently preserve, develop, and share our resources; all are called to excel in working with one another.

In the eternal present of the holistic self, outcomes in the ticking now of individual and team competitions are transient incidentals. This can seem nonsensical to the egoistic self when competing for personal superiority, as well as to the altruistic self when competing for social influence, but how could it be otherwise for the holistic self when seeking oneness with life? When that is our desire, our trust in the creative spirit of life is our joy and strength—its loss hopelessness.

For those who appreciate the call, life is an engaging climb filled with exhilarating vistas and troubling valleys as we become drawn ever upward by an inner refrain of a lot of little things being done well. Even when we're struggling through a valley of disappointments, an abiding joy and a satisfying peace come with the gathering awareness of how much of our world is a gift of the creative spirit. A beautiful expression of the trust and humility throughout that climb comes through in the African American spiritual "He's Got the Whole World in His Hands," sung by a voice that Italian maestro Arturo Toscanini said comes once in a hundred years, Marian Anderson's (https://www.youtube.com/watch?v=MRGZrv0Hl6k).

Appendix 2 is a look into the vistas and valleys that Walter and Margaret were glimpsing when writing this last volume of their letters. It's been noted how the expressions in this appendix consistently reflect their being drawn to the holistic spirits in Table 1. When looking at their expression in these discussions, we will be looking more closely at the nature of the personal concerns that gave rise to the guiding spirits they heeded.

Looking first at the universal concern for life, there were six broad categories of personal concerns: enjoying life, enjoying marriage, experiencing God, God's guidance, finding a fit, and the kingdom of God, each with its distinctive context. When it came to enjoying life, Margaret saw the possibility of their living like a king and queen if they could obtain a good kitchen set, a comfortable bed, and a piano (Appendix 2, 8/19). It would be years before Mom acquired that piano, but that valley of unmet material desires was not deep enough to keep her from later vocalizing and singing her thanks to God whenever she was hanging freshly washed clothes under a bright-blue Colorado sky. Walter was particularly mindful of having a Sabbath, a whole day set aside each week to rest from his work and to join with like-minded people in acknowledging the source of their lives and livelihoods (8/28).

When it came to enjoying their marriage, especially their first night together, Walter felt they needn't be worried if they were both mindful of the creative spirit that had brought them together (9/7). I wasn't there that first night, but less than a year later, Mary, my oldest sister, enriched their lives.

The contexts that bring God to mind can differ widely. Walter was mindful of the beauty of his world and the gift of his faculties (6/3). Margaret was mindful of her capacity to accept factors determining her being able to go on an exciting trip (6/6). Walter noted their sense of God's guidance while going about their daily work (6/15, 8/18). Margaret was also mindful of an inner awareness that could tell her when she had found the perfect song to sing for an occasion (8/25). Earlier, Walter shared an analogous inner awareness of a perfect fit when he wrote in his March 14 letter,

> As far as I have gone in seeking a helpmate, God has answered every prayer. Those tender and beautiful moments that every young couple experience are ours to look forward to and not to look back upon. Not only that, but with God's sanction and blessing, they will be

abiding and will not become timeworn.

Without God, this cannot be.

QGW, 169

Margaret's envisioning a vista of possibilities comes through in her mindfulness of God's promised kingdom coming to Earth and what that might mean for the expressed spirits she and Walter are birthing (7/5 and 9/10).

A spirit of mindfulness is key to experiencing the creative spirit of all that is. To get in touch with that sacred source and its gifts is to get in touch with your soul, the heart of your being. Should that be your desire, seek a more heightened awareness of what aspects of the creative spirit have brought a profound joy to our spiritual pathfinders but also to you. Ask that you be given the desire to feel a greater oneness with all of life.

Set aside a little time each day to briefly reflect on what heeding your mindful spirit might mean in how you go about your day. Find spiritual gatherings that give you a sense of being drawn into a oneness with life while speaking to your deepest questions, emotions, and curiosities.

Disclosure: Our Concern for Understanding

As noted earlier, our awareness of the elemental spirits outlined in Table 1 has been around a long time. We certainly see an analogous but instinctive awareness of the three spirits of disclosure in our pets and wildlife. The manipulating spirit of disclosure comes in multiple forms. An intimidating/boasting form is evident when male gorillas beat their chests to intimidate other males and strutting peacocks spread their trains to attract females. A suppressive form sends prairie dogs cowering into their holes when a hawk circles overhead. A deceiving form emerges in a cat stalking its prey and in the crippled walk of a killdeer with a "broken" wing that flies off once she has led you far enough away from her nest of eggs. We hear and see the consenting spirit of disclosure when birds noisily flock together prior to a migratory flight. We see and feel a truthful spirit on seeing a dog convey its happiness when greeting its owner. We hear it in a rooster's crow at the break of day but also in a rattlesnake's warning of its discomfort with our encroaching presence.

Our furred, scaled, and feathered friends have little choice when it comes to their disclosure instincts. We have a great choice. Like them, much of our awareness is shaped by what is seen through our own eyes, heard through our own ears, and felt with our own hands. Were that all, ours too would be a small world. It is not. Ours is a vast and rapidly expanding world that has been seen, heard, and felt by tens of billions of human eyes, ears, and hands throughout our history and has been shared over our relational web through the centuries. What we choose to disclose of our awareness of that world is limited only by our imaginations.

The vibrancy and enduring character of what is shared over that web, and consequently the world of which we are aware, is shaped by the elemental spirits of disclosure noted in Table 1. Those choices for egoistic and altruistic big selves basically reflect conceits advantageous to their most basic motivations in the ticking now—personal superiority and social influence.

When focused on personal superiority, the manipulating spirit of disclosure makes sense. Why give your counterpart an advantage by disclosing your knowledge and intentions when you can distract or deceive them? Why let them express themselves in ways that counter your intentions? However, there are downsides to this approach. Keeping one's understanding and intentions "under wraps" can be vexing. Suppressing the conceits of others can be fraught with difficulties. Ask any bully.

The consenting spirit of disclosure appeals to those focused on social influence. Its strength lies in its ability to profitably sway and be swayed. It creates the varied understandings and procedures of our different groups. It invites the sharing of thoughts and observations within a group. For those caught up in the process, it can be enlightening to hear what others have learned. It is gratifying when one is encouraged to speak. There are competitive downsides, though. It can be deflating when one is not encouraged to express oneself and discouraging when one is ignored.

The consenting spirit of disclosure has given rise to the special role the web of humankind now plays on Earth. Our schools, trades, businesses, organizations, and governments are the facilitating structures. Their effectiveness reflects the understanding, purposes, and practices that define and distinguish them. The purposes and practices are conceived and implemented by enterprising individuals heeding the consenting spirit of disclosure. The needed understandings are sourced in the scientific concepts that underlie our technical capabilities, the artistic creations that enrich our expressive behavior, the philosophical reasonings that underly our rules and conventions, and the religious teachings that can imbue all of them with life and meaning.

The institutions that formed around our need for trustworthy understandings are a consequence of the insistent heeding of the truthful spirit of disclosure by their founders and dedicated contributors. That spirit focused Galileo's and Darwin's scientific insistence on clarifying what could be objectively perceived through our senses. It radiated in Emily Dickinson's, Ludwig van Beethoven's, and Vincent van Gogh's artistic insistence on the enthralling expression of their inner worlds. It spurred Plato's and Aristotle's philosophical insistence on the role of reason in our broadest understandings. It anchored Moses's and Buddha's religious insistence on precepts freeing us from suffering and exploitation. Isaiah made clear that the disclosures of the truthful spirit would always be with us when he said, "The grass withers, the flower fades; but the word of our God will stand for ever" (Isa. 40:8).

All three spirits of disclosure were at play when that shot at Lexington and Concord was "heard round the world." The shot was a reaction to the manipulating spirit of the leaders of England in suppressing the colonists' freedom to vote on how they were being governed. Within a year, the signers of the Declaration of Independence unanimously proclaimed how they wanted their suppressors to see their reasons for firing that shot when stating, "We hold these truths to be self-evident, that all men are created equal, that they are endowed by their Creator with certain unalienable Rights, that among these are Life, Liberty and the pursuit of Happiness." Samuel Johnson truthfully wondered if those self-evident truths were being used for politically self-serving purposes when, in his 1775 essay, "Taxation no Tyranny," he asked, "How is it that we hear the loudest *yelps* of liberty among the drivers of negros?" (*GTWT*, 757). It would be almost another decade before a consenting spirit reigned broadly enough in the thirteen colonies to finally solidify voting privileges for most white men in the Constitution of the United States.

    Eighty years later, after a brutal civil war in which the divisive issue of slavery came to play a central role, the Fifteenth Amendment granted African American men the right to vote. It wasn't until my parents entered their teens that the Nineteenth Amendment finally granted women the right to vote. Walter cast those struggles as a fight for freedom when he wrote, in his October 30 letter,

> Freedom: Our nation fought the revolutionary war to gain religious and political freedom. We fought to free the slaves from their bondage. Theodore Roosevelt fought to free us from the bondage of great trusts being organized at that time. Women fought for the right to vote. Today they are fighting for equality with men

> in all things. We hear on all sides of us, concerning moral freedom of the younger generation.
>
> *EG*, 39

Unlike the "shot heard round the world," the bombarding of Fort Sumter shortly after the inauguration of President Lincoln triggered the armed conflict between the North and the South. The conflict was not notable for its personal reasons. Both sides liked the democratic rule of law sourced in the consenting spirit of disclosure, but they differed deeply in the type of material and human commerce that they wanted those rules to favor. The mob's storming of the United States Capitol on January 6, 2021, was notably different in its reason: to carry out the will of a president who had just been voted out of office. That such could occur in an established democracy requires an understanding of the power of the manipulating spirit of disclosure in the hands of a shrewd political leader who is driven by an entrenched desire for personal superiority and has the talent to sway crowds.

To understand the political and spiritual basis of this manipulative power, one must bear in mind three points: First, a predominantly self-aggrandizing person will shove aside any rule of law if doing so advances their inherent egoistic desire for personal superiority—that's the "law of the jungle" that Spinoza called the natural state. Second, a political leader is powerless without a team of like-minded colleagues and a large group of motivated supporters.

Third, most humans are altruistically oriented when not threatened. We are attracted to the promoting and consenting spirits in the groups with which we identify and, consequently, feel that the rules of a game should be followed in team competitions.

Everyone knew the outcome of the voting tally for the presidency in the 2020 election would be close. Both parties were going to great lengths to assure that they were not being treated unfairly in the election. What was the then-president to do when the final tally of the votes said that he had lost? Because the vast majority of his supporters, like the rest of us, feel we must abide by the rules of the games we play, he would lose their support if he didn't follow the rules of the game and concede—unless he could convince them that the voting was rigged. "The election was stolen" became his battle cry. It became known as "The Big Lie" in the nonconservative news media.

On January 6, 2021, it was the vice president's responsibility to open the box of electoral votes from each state and have them officially tallied in a joint session of Congress. Before the box was to be opened, the then president of the United States had convinced a mob of his supporters chanting "Hang Mike Pence" to break through the protective gates and surrounding security forces in order to thwart the long-honored peaceful transfer of executive power—a hallmark of democratic rule of law. (Some of his enablers would later say the phrase "Hang Mike Pence" should be interpreted "metaphorically," a nuanced interpretation easily lost in the heat of a conflict in which our congressional women and men sought protection behind and under their desks before they were hustled by security staff through a side door to a more protected area.)

A legal judgment of what happened on January 6 is now being made. Under a government based on the rule of law, such a judgment must reflect an expert's determination of the specific laws that were violated by the assailants of the Capitol and their enablers. The book *Oath and Honor: A Memoir and a Warning* by Liz Cheney (a leading conservative voice of the Republican Party at the time) is possibly the best insider revelation of the power of the manipulating spirit of disclosure at work during the assault and during her essential service on the select committee as it attempted to get an overall and reasonably objective grasp of what transpired on January 6. Our spiritual judgments of what happened on January 6 concern us here.

We can only imagine what those personal judgments will be for the assailants, but imagine we must, lest we find ourselves enabling and participating in a comparable situation later in life and wishing we hadn't. The assailants surely gave much thought to what their embodied spirits wanted to do when they stormed the Capitol. But did they give any thought to the nature of the spiritual seeds their leader was getting them to sow into the expressed spirits they were birthing and the world would be judging? What seeds were they sowing for all the world to see when wearing their MAGA hats as they attacked the legislative body of our democracy? Weren't the assailants doing something most of them would once have thought appalling? Weren't they supporting the effort of an egoistic leader with an overweening desire for personal superiority trying to create a

"MAGA" party with loyal enablers subservient to his will? What will be their experience when confronted at their death by the judging spirit of their and my Lord, who, in the last quote under the disclosure section in Appendix 3, says, "The devil . . . is a liar and the father of lies."

Heeding the spirit of truthful disclosure was a simple matter for Margaret and Walter, even when it came to money. On February 5, Margaret writes, "I don't believe that I have told you what my salary is. I get one hundred dollars a month. I put thirty dollars in [a] savings account" (*QGW*, 5).

That same day, Walter writes,

> Were money to be my goal in life and my great desire, I know that by pursuing the right course I could amass a fortune, but that does not seem right to me. Does it to you? Now I want to provide for you, to make you comfortable and happy, but I am going to take God at His word, and He has said, "But seek ye first his kingdom and his righteousness; and all these things shall be added unto you."

QGW, 13

On March 30, Margaret writes, "Walter, I don't think we should need so much money to live on. We wouldn't need to have such expensive furniture" (*QGW*, 262). On March 31, Walter writes,

> Although I stand to lose about 150 dollars on my cattle, yet they pay all I owe the bank with a little left over. My potatoes will help pay all other debts, and I believe I will have a little money left for running expenses."
>
> *QGW*, 265

There is no scheming, no getting the "raw end of the deal," no feeling of being cheated or outsmarted. There is only the assurance of each having the other's interests at heart.

The spirit of truth is central to the vibrancy and freedom Margaret and Walter have found in each other and that they anticipate in marriage. After their Christmas commitment to each other, they expressed their interest in openly sharing their desire for and excitement in finding a partner in whom they could share their mindfulness of the creative spirit. In his January 16 letter (*EG*, 129), Walter writes, "I can hardly believe that God has directed me . . . to whom I can write and talk to about spiritual things." In her January 22 letter (*EG*, 144), Margaret responds, "Walter, I wouldn't enjoy writing letters if I couldn't write anything spiritual because that is the only thing that is lasting through time and eternity."

There are six categories of personal concerns that arose in this third volume in which disclosure seemed the underlying concern:

abiding in God, connectedness, experiencing God, helping others, finding a fit, and the kingdom of God. Margaret heeds a consenting spirit of disclosure in two of the thirteen listed concerns. We never see them heeding a manipulating spirit of disclosure.

Margaret's consenting spirit of truth comes through in writing that "Jesus' body had to be cut" in order that we might "bear more fruit" (6/12); her truthful spirit comes across when she describes the out-of-the-blue helpfulness of her connectedness within the relational web (5/17 and 7/13). Her consenting spirit appears again in her seeing the "power of prayer" when a movie theater is shut down (5/31).

Walter's truthful spirit comes through in his trying to understand Leonard's conversion (6/15) and in his conveying the role verses of the Bible play in our world (7/14). It also appears in the help he feels is brought to others in the relational web via listening (5/16) and encouragement (7/21).

Both heed the truthful spirit when it comes to their inner peace: Walter in contemplating God's promise and nearness (5/12) and Margaret in a song that rings in her ear (8/15). They also heed it when it comes to their relationship: Margaret in anticipating Walter's care for her (5/20) and Walter in seeking help from God in resolving their disagreements (5/20) and in understanding Margaret's love (6/19).

A spirit of truthful disclosure grounds our deepest understanding. That spirit is latent in all of us and finds its fullest realization in those at one with all of life. Should you desire it, pray for clarity in your truth, the courage and love to share it kindly, and the strength to persevere. Along the way, seek exchanges with others in which the truthful spirit reigns.

Worthiness: Our Concern for Well-Being
Worthiness is a multisourced concern in every mind. It comes with entitlements and attendant constraints that are quite differently perceived and appreciated by our egoistic, altruistic, and holistic self-orientations.

The ranking spirit of the egoistic big self is steeped in our evolutionary past (Spinoza's native state). Pecking orders are a common occurrence among animals. I've watched dominant young chickens feel entitled to peck others they've intimidated whenever their heads come within reach at a feeder.

Though lacking the literal beak, we have our own pecking behaviors: cutting remarks, judgmental comments, exclusions, and bullying, to name a few. Jesus took a dim view of these expressions of the ranking spirit when he said, "Everyone who is angry with his brother shall be liable to judgment; whoever insults his brother shall be liable to the council, and whoever says, 'You fool!' shall be liable to the hell of fire" (Matt. 5:22).

We can imagine but cannot know how the chickens at the top of the pecking order experience their freedoms and how those at the bottom experience their fears. Nor can we comprehend the experiential disparities between those at the top of the current human pecking order who feel entitled to mega-million-dollar yachts and estates and those at the bottom who beg in our streets. We do, however, know how a cutting remark can hurt. If we were aware of their effect on the expressed spirit we eventually birth, we would shy away from them.

In *A Letter Concerning Toleration*, John Locke drew our attention to the entitlements that can come with obeying the rules and conventions of a commonwealth:

> The commonwealth seems to me to be a society of men constituted only for the procuring, preserving, and advancing their own civil interests. Civil interests I call life, liberty, health, and indolency of body; and the possessions of outward things, such as money, lands, houses, furniture, and the like.

GTWT, 690

The grouping spirit of the altruistic self is central and common in any commonwealth. That spirit underlies our broad acceptance of the roles and modes of conduct that come with the development of the groups with which we identify and in which we participate. We have families, communities, religious affiliations, and countries into which we are born and live. We have associations to which we are drawn by our interests, talents, and training. In addition to the meaning they can bring to our lives, these associations foster a worthiness grounded in our perceived character of the group.

Individuals in the group in which we participate are often the first recipients of the words and actions that seed our expressed spirits. Being sympathetic interpreters, they are likely to share the spirit and may even recollect and pass on a typifying expression in which it was embodied. They are the "wombs" involved in the birthing of our expressed spirits. Consequently, we have a natural interest in promoting the worthiness of all members of our interest groups. Paul enjoins those appreciating his teachings to encourage the worthiness of others when he tells the Philippians in first-century Greece, "Do nothing from selfishness or conceit, but in humility count others better than yourselves. Let each of you look not only to his own interests, but also to the interests of others" (Phil. 2:3–4).

Much of our joy in life comes about through the diverse roles we play. The grouping spirit brings to each role a sense of mutual worthiness that is often undermined by the ranking spirit when the material rewards linked to those roles are more important than the joy that comes with doing our part. The encompassing spirit of worthiness recognizes the value of all the roles that must be played. That was emphasized by Paul when he told the Christians in Corinth,

> The eye cannot say to the hand, "I don't need you!" And the head cannot say to the feet, "I don't need you!" . . . God has put the body together . . . so that there should be no division in the body, but that its parts should have equal concern for each other. If one part suffers,

> every part suffers with it; if one part is
> honored, every part rejoices with it
>
> 1 Cor. 12:21–26, NIV

Self-worth knows no genetic, national, ethnic, or bodily boundaries. It is a gift of birth. It comes with being a child of the creative spirit. It comes with experiencing and insisting on the little self. It is a worthiness that can be obscured by a misplaced emphasis on the competitive nature of the big self. Paul addressed the problem roughly a decade later when he said to the Colossians,

> Put to death, therefore, whatever belongs to
> your earthly nature: sexual immorality,
> impurity, lust, evil desires and greed . . . Do not
> lie to each other, since you have taken off your
> old self with its practices and have put on the
> new self, which is being renewed in knowledge
> in the image of its Creator. Here there is no
> Gentile or Jew, circumcised or uncircumcised,
> barbarian, Scythian, slave or free, but Christ is
> all, and is in all.
>
> Col. 3:5–11, NIV

There are stark differences between wholesome and divisive competition. Wholesome competition reigns when we emphasize the mutual joy in our social endeavors. We benefit from the play of individuals in interest groups, athletic teams in leagues, ethnic groups in communities, political parties in countries, companies in economic enterprises, and countries in the global commonwealth. That collective joy is a gift of the encompassing spirit of worthiness.

Divisive competition arises when we emphasize winning in those endeavors. The encompassing spirit of worthiness wanes, families split, leagues fold, suppression spreads, and wars break out. In the case of politics, it may only take an unsettling time and an opportunistically gifted demagogue with a ranking spirit of worthiness who can emotionally elevate her or his followers by trampling on the consensual rules of politics. As a result, these followers will question the loyalties of competitors and castigate as "outsiders" those they are prone to blame and can be taught to hate. The consequences can be great and sometimes unimaginable.

That happened in 1933 when Hitler was voted into power in the Weimar Republic, "widely hailed [as] the most modern democratic constitution of its day" (*Britannica*, 2005, 20, 118). His self-aggrandizing appeals to Aryan supremacy ignited his followers and attracted sympathizers. He blamed the Jews and Communists for Germany's defeat in the Great War by saying they had undermined the ideal of Germanic supremacy. His degrading tirades and the intimidating tactics of his brown-shirted storm troopers encouraged and freed the anger of his supporters. When a teenage Polish Jew shot a German official, he found a "justifiable" opportunity to unleash that anger. Two days later, on what is now called Kristallnacht, the Nazis broke into and plundered Jewish businesses and synagogues across

Germany, Austria, and the Sudetenland. This fed the desires of the Nazis for the rewards of "social superiority." They willingly created a police state and went to war per Hitler's urging them "to obtain by the German sword sod for the German plow and daily bread for the nation" (*Mein Kampf*, translated by Ralph Meinheim, Kindle, 140).

The spirit he unleashed throughout Germany triggered the Holocaust we associate with his life and the lives of his supporters. We may portray—but cannot grasp—the unconscionable starvation and gassing of millions of Jews by the Nazis, who were once their neighbors. We may find ourselves aghast at the orders more of men, women, and children who were crippled and killed as a consequence of the world war the Nazis perpetrated, but we cannot fathom the scale of the physical and psychological pain of all the families who suffered. Can we even think of the expressed spirits of the perpetrators without smelling the stench of mass death, seeing the starved bodies piled on top of each other, hearing the point-blank gunning of the helpless captives they had rounded up, or feeling the barbed wire tearing the flesh of prisoners trying to escape? Once Hitler found a receptive ear for his notion of Aryan supremacy, his unswerving supporters seemingly became oblivious to the gangrenous effect his exclusive promises were having on the expressed spirits they were birthing.

In any commonwealth of individuals competing for measurable rewards, most will be left anxious, and many will be left worrying for the well-being of their families and friends. In Matthew 6:31–34, Jesus contrasts the big self's anxiety about material worth with the little self's encompassing spirit of self-worthiness when he says,

> Do not be anxious, saying, "What shall we eat?" or "What shall we drink?" or "What shall we wear?" For the Gentiles seek all these things; and your heavenly Father knows that you need them all. But seek first his kingdom and his righteousness, and all these things shall be yours as well. Therefore do not be anxious about tomorrow, for tomorrow will be anxious for itself. Let the day's own trouble be sufficient for the day.

In Matthew 22:37–40, when asked what one must do to abide in the kingdom of his father, Jesus drew attention to the encompassing spirit of worthiness that unites us all when he said,

> You shall love the Lord your God with all your heart, and with all your soul, and with all your mind. This is the great and first commandment. And a second is like it, you shall love your neighbor as yourself. On these two commandments depend all the law and the prophets.

Mohandas Gandhi expressed this sentiment another way when, in his desire for oneness with life, he holistically linked the spirits of self-worth and truth in writing, "To see the universal and all-pervading Spirit of Truth face to face one must be able to love the meanest of creation as oneself." Similarly, Rabbi Hillel enigmatically captured the engulfing self-insistence of the encompassing spirit when saying, shortly before the birth of Jesus, "If I am not for myself who is for me? And in being for mine own self what am I? And if not now, when?"

When heeding the encompassing spirit of worthiness, we can experience the joy of interacting with anyone who appreciates our seeing them as a wanted member of the web of humanity. We see that joy in the courtship letters of Margaret and Walter when she befriends a fellow teacher who felt men had a low opinion of her (Appendix 2, 5/23) and when Walter speaks kindly of his alcoholic landlord (7/25).

The more mindful we are of the power of the creative spirit within us, the greater our awareness of our rights as children of that spirit and our strength to exercise them. Both are needed when reproving the in-vogue behavior of a group of colleagues that violates a moral injunction of your heart. Margaret was struggling with the issue when, in her November 27 letter, she wrote, "I can't understand why people can't be interested in other things than cards, dances, and movies. There would be so many things I would enjoy doing if I had the time" (*EG*, 72). She expressed the consequences of her reproofs when she wrote in her January 16 letter, "There was a bridge party last night. . . . I am glad that I wasn't asked" (*EG*, 126) and in her March 21 letter, "Last Thursday the juniors had a dancing party. They invited all the teachers except me. I think they were very considerate of my wishes" (*QGW*, 185).

When Walter wrote in his December 5 letter (*EG*, 83), "There is much lust in us, lust of pride, lust of money, of security and lust for the gilded happiness offered by the devil," he didn't mention the lust for ethnic and racial superiority that the spirits of group and rank worthiness can engender in the unmindful. Yet when these letters were being written, Hitler was successfully promoting those desires in trying to "make Germany great" with promises of Aryan supremacy. Not all Christians in Germany were caught up in carrying out Hitler's will. There were some whose encompassing spirits of worthiness befriended Jews. They were threatened, ostracized, jailed, or sent to concentration camps. Many gave their lives knowing their expressed spirits would live on, notably like that of the theologian Dietrich Bonhoeffer.

The encompassing spirit and its inherent strength are latent in all of us. Should you desire it, reserve a time each day to pray for clarity in understanding its transcendent nature and how your self-worth figures into your joy in this life and the life hereafter. Then spend a moment contemplating how the ranking, grouping, and encompassing approaches to worthiness shape the nature of our empathy for one another.

Relationships: Our Concern for Others

Early in childhood, we learn to recognize the three elemental spirits that arise when developing our relationships with others: serving, role-playing, and dominating. We often like to dominate when competing in games and fighting over toys. We enjoy playing roles in pretend play. We serve by sharing our toys, teaching others our games, and giving all a chance to play.

As we grow older, time frames lengthen, interests evolve, and concerns deepen. We learn that team performance is more important than individual play and that the game is more important than the play of the team. That's how things are supposed to be: game first, then team performance, then individual play. That's how things have happened through the centuries and will continue to shape our future. When the game is more important than winning and teamwork is more important than standing out individually, harmony reigns. But things can, and too often do, fall apart. When they do, it is important to see why a misplaced emphasis among our elemental spirits is so often involved.

We often weigh the pros and cons of our choices, both short term and long term, when addressing our concerns for life, disclosure, and worthiness. Yet most of the time, we don't give much conscious thought to the spirits we heed when it comes to relational concerns. We can't, and for good reason. Even the simplest conversation or social activity could entail numerous relational concerns, each potentially begetting a freedom to serve, need to carry out a role, or desire to dominate. We would go raving mad if we had to grapple with the short- and long-term consequences of the complex interplay of those relational choices. Moreover, if our different senses of self (holistic, altruistic, and egoistic) were to change with every relational concern arising in our conversations, we would be seen as rationally and emotionally incoherent. Instead, during each interactive encounter, we heed the choices to which our self-orientation is habitually drawn.

It is helpful to imagine how the serving, role-playing, and dominating spirits might play out in a hopefully lasting relationship, such as the spousal relationship anticipated in these letters. The letters have suggested many reasons such a relationship might come about: companionship, sex, having children, enjoying couple get-togethers, an enriching home life, shared workloads, et cetera. Each grants an additional type of meaning and worthiness to the lives of the participants.

In a spousal relationship between two individuals seeing each other and their worlds through the eyes of the big self, competitive comparisons are inevitable. That's the nature of the big self. Whether the comparison is viewed positively or negatively, explanations will be sought. When both spouses view their relationship through egoistic eyes, they will naturally gravitate to "I" explanations when the judgments are positive and to "you" explanations when they're negative. When negative judgments predominate, their dominating spirits can lead to verbal, emotional, and physical conflicts with potentially harmful consequences.

That is less likely to happen when both spouses view their relationship through altruistic eyes. That orientation will always encourage them to abide by and set up rules and roles by which they can get along. Although their competitive big-self natures will evoke both positive and negative comparisons with other marriages or families, both will naturally gravitate to "we" explanations. Problems can still arise when an individual in the marital pair envies the joy others are seemingly finding in their spousal relationships. Their role-playing spirits can improve matters by better sorting out who does what and when. However, such agreements are meaningless should they start questioning the worthiness of their spouses.

That the encompassing spirit of worthiness is central to appreciating the inherent self-worth of one's marital partner was recognized long ago. It is well described in *The Inclusive Bible: The First Egalitarian Translation* in Genesis 2:18–24:

> Then YHWH [the Hebrew spelling of God] said, "It is not good for the earth creature to be alone. I will make a fitting companion for it." So from the soil YHWH formed all the wild beasts and all the birds of the air . . . but none of them proved to be a fitting companion, so YHWH made the earth creature fall into a deep sleep, and while it slept, God divided the earth creature in two, then closed up the flesh from its side. YHWH then fashioned the two halves

into male and female, and presented them to one another.

When the male realized what had happened, he exclaimed,

This time, this is the one!

Bone of my bone and flesh of my flesh!

Now, she will be called Woman, and I will be Man,

Because we are of one flesh!

This is why people leave their parents and become bonded to one another, and the two become one flesh.

Nothing is being said about the rank or group worthiness of either of the individuals, nor is anything being said about sex or even children, although those possibilities are certainly implied. Rather, the writer is conveying the joy that can be found in entering a time-honored and enduring relationship with another human being who complements us. That complementarity is a merger of two neural relational webs in which full and complete intercourse is both a desire and a gift. It is helpful to briefly look at the serving, role-playing, and dominating spirits in some of our other relationships before illustrating the holistic spirit in the emerging partnership between Margaret and Walter.

Our broad roles of provision, protection, and parenthood have been around a long time. Although these are largely shared by plants and other animals in one form or another, our altruistic tendencies have increasingly set us apart. In *Moral Origins* (page 67), Christopher Boehm claims that "45,000 years ago very likely almost all humans on this planet were practicing . . . egalitarianism." These tendencies have given rise to the vast distinctions in the existentially interdependent roles we now play in the human relational web.

The role-playing spirit of the altruistic self is central. We have duties vis-à-vis our children. We are supposed to carry out our study assignments in school. We're expected to participate in social groups. We're urged to enter vocations in which we can provide for ourselves and families. We may be obligated to serve militarily and possibly urged to serve politically. When there is cooperative and jointly appreciated team play, we all benefit materially and corporately.

Such play isn't always the case. Walter felt the "gilded happiness offered by the devil" can lead us astray. Over three thousand years ago, as described in Exodus 20:13–17, we were enjoined from falling prey to an overzealous heeding of the dominating, manipulating, and rank-ordering spirits of the egoistic self:

> You shall not murder.
>
> You shall not commit adultery.
>
> You shall not steal.

> You shall not bear false witness against your neighbor.
>
> You shall not covet your neighbor's house; you shall not covet your neighbor's wife, or male or female slave, or ox, or donkey, or anything that belongs to your neighbor.

Although the first four of these commandments appeal to the role-playing spirit of the altruistic self, they mean little when we are egoistically self-orientated and so situated that we don't fear getting caught or we see ourselves above the law.

The coveting commandment is more problematic. Carrying out our roles can be deeply satisfying and meaningful when we are altruistically oriented, but should competition for social influence dominate our time and attention, all can be lost. Commitments and obligations can mount to the extent that we lose our joy in life and in others. Paul notes the holistic answer to this issue in his letter to the Corinthians:

> If I speak in the tongues of men or of angels, but have not love, I am only a resounding gong or a clanging cymbal. . . . If I have the gift of prophecy and can fathom all mysteries and all knowledge . . . if I have a faith that can move mountains . . . if I give all I possess to the poor and give over my body to hardship that I may

> boast, but do not have love, I gain nothing. . . . / Love never fails. But where there are prophecies, they will cease; where there are tongues, they will be stilled; where there is knowledge, it will pass away. . . . / These three remain: faith, hope and love. But the greatest of these is love.
>
> 1 Cor. 13:1–13, NIV

Although interest in personal superiority and social influence can drive our will to excel, our desire for oneness with life can be stronger and, when so, certainly more enduring. Lasting joys come when developing and using our talents to further our lives and the lives of those we love. Heeding the serving spirit in our daily affairs and vocations brings joy to others. Not only does their joy impart joy and meaning to our own lives, but it also enhances our vistas of life in the hereafter.

We see that serving spirit in parents who love their children, in teachers and mentors who appreciate their students and trainees, in salespersons who value their customers, in employers who care for their workers, and in leaders who respect the lives of their followers. It is publicly declared whenever we take oaths to serve in some capacity, as Walter and Margaret did when they met at the marriage altar.

Appendix 2 lists five types of involvements for Walter and Margaret in which the underlying concern was relationships: enjoying life, enjoying marriage, helping each other, helping others, and their relationship. When it came to enjoying life, Walter felt their tasks would be lighter if seen in light of a higher purpose, clear evidence of a serving more than a role-playing spirit (Appendix 2, 6/14). His anticipating the joy in the study and exchange of viewpoints with Margaret does so as well (7/12), as does his not wanting Margaret to cut her trip short (6/28). His freedom from being locked into a conventional spousal role came through when he wrote in his August 18 letter, "Yesterday as I said, I scrubbed two of the rooms. Perhaps I should have said nothing of this ability. Oh well, it will be fun to help you." That holistic understanding of service is also seen in his wanting their house to be the way Margaret envisioned it (7/8). While wondering "about that which we call love," he concluded that it was expressed in "service to one another" (6/28). Margaret is glad her words are helping Walter (5/11); in turn, he is looking forward to helping and encouraging her (5/17).

Their interest and joy in life and their love for each other continued throughout their marriage. Both were apparent to me even as a teenager, but I may never have benefitted by seeing in these letters the needed understanding had not my priorities been upended in the Night of the Little Self.

The spirit of service, so central to all enduring relationships, is latent in all of us. There is a deep joy that comes in seeing and helping others find joy in using their talents. That relational reward comes when you feel God's love and others feel God's love through you.

Should that be your desire, ask that you might receive the gift of an encompassing spirit of worthiness that frees your serving spirit to reach out to whomever the creative spirit brings your way. Then if you pray for a truthful and consensual spirit of disclosure that can hear and appreciate your true needs and those of others, opportunities for your serving spirit will open up in your family, friends, community, and throughout your world.

Brokenness: Our Concern for Rectification
Brokenness is a universal concern. It is frequent and endless in its reasons and can range from personal to global. Whatever the case, we have the choice of addressing it egoistically by finding fault outside ourselves, altruistically by seeking the help of others, or holistically by searching for our own responsibility for the problem. All are important.

When worried about a possible second world war, Margaret wrote, in her November 11 letter,

> I can't understand why people are unable to live peaceably together but then I realize that I get angry and say cutting remarks that can cause hurt feelings so it is only God can help humanity thru the peace of Christ to live harmoniously together.

EG, 53

That same day, Walter wrote, "Our statesmen, seemingly rack their brains for a way to promote peace. I wonder if it is a confidence, in our own ability which causes us to overlook the simple way that our savior offered" (*EG*, 56).

Centuries before Jesus proclaimed the peace that Margaret and Walter were seeking, Ezekiel (34:2–4) noted the problem that can arise between leaders and those they lead:

> "Son of man, prophesy against the shepherds of Israel; prophesy and say to them: 'This is what the Sovereign Lord says: Woe to you shepherds of Israel who only take care of yourselves! Should not shepherds take care of the flock? You eat the curds, clothe yourselves with the wool and slaughter the choice animals, but you do not take care of the flock. You have not strengthened the weak or healed the sick or bound up the injured. You have not brought back the strays or searched for the lost. You have ruled them harshly and brutally.'"

In *Doing Christian Ethics from the Margins*, Miguel A. De La Torre, a professor of social ethics, gives a current take on the daunting moral task before us. He writes, "The ethical task before both those who are oppressed and those who are privileged by the present institutionalized structures is . . . to dismantle the very structures responsible for causing injustices along race, class, gender, and orientation lines." Reinhold Niebuhr, a prominent Protestant theologian, recognized the spiritual difficulty when he wrote in *Moral Man and Immoral Society*, roughly twenty years before the civil rights movement, "Society accepts injustice because it does not analyze the pretensions made by the powerful and privileged groups of society."

The joy of team sports suggests what could be between peoples of different cultures, races, beliefs, and professions. Players broadly accept the rules of play, coaches encourage team play, and overseers (referees, promoters, administrators) promote the game by shaping and enforcing the league rules that assure interest in the sport. The best players are skilled in heeding their egoistic desire to win in moments of one-on-one competition, recognizing opportunities for cooperation in moments of team play, and sharing their knowledge of the game with teammates, students, and children. The best coaches are adept in teaching the rules of the game and the spirit of effective team play but also in recognizing the importance of games in which the outcome is unknown until the very end. The best overseers are proficient in conveying the importance of fair play and even playing fields, of having creative rules, and bringing more into the game via new leagues and venues for the differently talented and circumstanced. In doing this, individually competitive spirits are recognized and appreciated, but they are subordinated to the holistic spirits that emphasize fair and rewarding opportunities for all involved.

In the game of life, we have competing constituencies and parties instead of teams, and we have political leaders instead of coaches and overseers. Given those parameters, there will be competition for resources and opportunity, as there is in sports. But if we are to have wholesome competition, there must be a widespread desire for inclusive participation, good rules, fair play, even playing fields, and impartial refereeing, just as there is in any healthy sport. Our individual interests must be subordinated to the interests of our defining constituencies, and they, in turn, must be subordinated to the enduring interests of humankind. That will require "overseers" who can elevate the holistic spirits of the little self over the altruistic and, especially, the egoistic, spirits of the big self.

As is readily apparent in the aforementioned books by Torre and Niebuhr, we are far from the unity and peace we seek. In sports, there is a clue for a critical means by which that global brokenness can be addressed. Brokenness in sports arises when teams consistently don't perform well and there is structural unfairness between competing teams. The first type is remedied by replacing the responsible coaches and the second by replacing the responsible overseers.

But how do you replace leaders who forcefully oppose their removal? It's been done by natural death and coups d'état in many cases; by revolution in America, France, Russia, and China in the past two and a half centuries; and by passive takeovers led by Mohandas Gandhi in India, Ruhollah Khomeini in Iran, Boris Yeltsin in Russia, and Nelson Mandela in South Africa in the past century. But it is difficult—especially so when dealing with self-aggrandizing leaders willing to violate established rules just to stay in power.

Brokenness becomes personal when we dislike what is going on in our daily lives. Its inevitability is reflected in the down-the-hill setbacks in the big-self view of life and in the troubling valleys in the little-self view. Faultfinding, help-seeking, and self-searching all come into play, as is easily seen in my parents' courtship letters.

In her July 19 letter, Margaret's faultfinding spirit noted an unwanted swelling on her foot. Her altruistic, help-seeking spirit went to work and directed her to a nurse who, in turn, suggested she seek additional help from her doctor. In her July 23 letter, she mentioned Dr. Will's surgically removing the swelling but advising her to stay off her foot for possibly three weeks. She had hoped to visit Walter and discuss the changes he was making to their new home. When she realized that might not be possible until after their anticipated marriage in September, her self-searching spirit soon brought back a joy that led her to write, "Now I have additional time for meditation; God knew my need" (Appendix 2, 7/23).

It would take Walter a month and a half to get his mind around the financial disappointment he had suffered when he wrote, in his March 31 letter (*QGW*, 265), "For I have sold my cattle, and tho my prayer was that I might make money on them, yet I always asked that I might be given the grace to say, Thy Will be done." His self-searching spirit was evident when he wrote, "Do you suppose God uses this way to help me learn how to take care of money and myself? . . . Perhaps God is granting just enough to supply my need."

Further on, his help-seeking spirit put a mistaken faith in the law of averages when he writes, "I have lost two times on cattle. The law of averages says I win next time" and hoped Margaret would concur. His help-seeking spirit attempted another consolation in his April 3 letter when he wrote, "Tho I lost some money in my farming & feeding, others have lost much more." On April 6, his self-searching and truthful spirits synergistically merged when he wrote, "I wanted the cattle and potatoes to bring me 5 or 6 hundred dollars. Then I would have thot that God surely means this so that Margaret and I can be married right away."

We see his prayer for grace answered and his joy in farming renewed when he writes in his May 17 letter,

> This money question need not worry us at all. I am not going to worry about it anymore. Our needs will be adequately cared for. I shall work and take every possible care of all my crops and then take what I receive.

Two months later his self-searching spirit is still seeking and possibly finding meaning in his financial disappointment (7/24).

The faultfinding spirit is more common and naturally heeded when addressing relational brokenness. That spirit emerges when Margaret writes in her March 12 letter (*QGW*, 160), "I haven't written before because I was waiting for a letter from you. It came last night. . . . Walter, if you . . . don't want to get married next fall, I wish you would say so." Later, her self-searching spirit reveals her mistaken interpretation of Walter's letters. Two days later, she heeds her help-seeking spirit in writing, "I want you to forgive me for what I wrote."

Walter's faultfinding spirit recognized that Margaret has misread what he was trying to say. His egoistic faultfinding spirit could have said, "You should read more carefully!" and his altruistic help-seeking spirit could have said, "We should read more carefully!" Instead, his holistic self-searching spirit sought out his responsibility in the problem, and he wrote, on March 17,

> You had good reason to believe as you did, and
> the fault was mine. I am so glad that you took
> it to God and that He revealed the truth to you.
> It was so good to know that God answered my
> prayer for peace of mind and heart for you.
> Margaret, you need never fear in that respect
> again. You are the only girl I want, and each
> letter I get adds to my certainty.

Walter's faultfinding spirit came to the fore when he lost his joy on seeing how his fellow Americans celebrated their "privilege of religious freedom." His self-searching spirit helped him recognize that he needn't let his faultfinding spirit keep him from celebrating that privilege in a suitable manner (7/4). Margaret also lost her joy when she gave her faultfinding spirit free rein to focus on her "failures and short coming." But as soon as her self-searching spirit "emptied her heart of sin on Christ and looked up to Him," her joy returned (9/2)—an expression that reveals both the power of the self-searching spirit and consenting spirit of disclosure.

Rachel Naomi Remen's book *Kitchen Table Wisdom: Stories That Heal* is filled with heart-gripping cases in which wisdom and understanding come through in struggles with brokenness. In one case, she writes,

> A young man with osteogenic sarcoma of the right leg . . . had been a high school and college athlete and until the time of his diagnosis his life had been good. . . . Two weeks after his diagnosis . . . they had removed his right leg above the knee.

In writing, "He refused to return to school. He began to drink heavily, to use drugs, to alienate his former friends, and to have one automobile accident after another," she noted how that young man was heeding his angry faultfinding spirit.

In recognizing the young man's need for his help-seeking spirit, his former coach convinced him to talk with Dr. Remen. She reveals the man's awareness of his truthful spirit of disclosure when she writes,

> In our second meeting . . . I gave him a drawing pad and asked him to draw a picture of his body. He drew a crude sketch of a vase, just an outline. Running through the center of it he

> drew a deep crack. He went over and over the crack with a black crayon, gritting his teeth and ripping the paper. . . . It hurt to watch. After he left, I folded the paper and saved it.

Dr. Remen goes on to describe one of the endlessly different and fascinating ways in which the self-searching spirit can address our concerns for brokenness. She writes,

> In time, his anger began to change in subtle ways. He began one session by handing me an item . . . about a motorcycle accident in which a young man had lost a leg. His doctors were quoted at length. I finished reading it and looked up. "Those idiots don't know the first thing about it," he said furiously. Over the next month he brought in more of these articles . . . His reactions were always the same . . . No one understood them, no one really knew how to help them. . . . I asked him if he wanted to do anything about it. Caught by surprise, at first he said no. But just before he left he asked if I thought he could meet some of these others who suffered injuries like his.

She goes on to write,

Within a few weeks, he had begun to visit young people on the surgical wards whose injuries were similar to his own. . . . The surgeons, delighted with the results, referred more and more people to him. . . . As he got to know them, his respect for them grew. Gradually his anger faded and he developed a sort of ministry. I just watched and listened and appreciated. . . . In our final meeting, we were reviewing the way he had come . . . I opened his chart and found the picture of the broken vase that he had drawn two years before. . . . He took it in his hands and looked at it for some time. "You know," he said, "it's really not finished." Surprised, I extended my basket of crayons to him. Taking a crayon, he began to draw lines radiating from the crack in the vase to the very edges of the paper. Thick yellow lines. I watched puzzled. He was smiling. Finally he put his finger on the crack, looked at me, and said softly, "This is where the light comes through."

A spirit of self-searching is key to finding your responsibility when problems come your way. If you desire that spirit, ask how the creative spirit might use you in envisioning and becoming an instrument of the needed change. If the problem is your joy in life, set aside a quiet time and find a secluded place where you can let that "still, small" voice eventually say where your true joy lies. Once that is known, your most trusted guiding spirits will lead you there.

Death: Our Concern for What Lives on at the End of Our Lives
The last ultimate concern is our inevitable death. The guiding spirits that emerge when we're faced with that concern are judgmental in their nature. The appreciation of the attending joy or sorrow rests on the distinction between the materiality of our embodied spirits and that of our expressed spirits. As noted earlier, the materiality of the former resides in us and basically shapes our will and judgment. The materiality of the latter resides largely in the minds of others and is variously reflected and dispersed in their dispositions, joys, and judgments.

The latter materiality is seeded by our expressions: the comments we do or don't make, the actions we do or don't take, and the creations we do or don't promote. These expressions radiate throughout our lives. Their circumstances and material particularities are continuous and daunting and, for the most part, quickly fade from the recipient's mind. The spirits they convey—the elemental spirits we have been discussing—are different. They are few and ever recurring. Consequently, we must look at the spirits we have been heeding to catch a glimpse of how our expressed spirits might be judged as they go out into the world.

That judgment at death's door could be harsh should our habitual orientation be egoistic. The dying spirit of that orientation will offer pain rather than hope. The physical presence that may have meant so much earlier will mean little, the desire for personal superiority even less. The benefits of material acquisitions inevitably fade as their ownership moves on, possibly to those our dominating spirits may have tried to better. That dying spirit will likely revive the pain of past consequences of heeding our self-aggrandizing, manipulating, ranking, dominating, and faultfinding spirits. That pain, when unforgiven, doesn't go away.

As a fearful harbinger of what is to come, the dying spirit has a corrective role to play. It can alert us to the need to drop any overriding desire for personal superiority before it is too late. That happened for John Newton when he commanded a slave ship bound for America. Years later, as an Anglican priest, he poetically captured in "Amazing Grace" his joy in experiencing a fearful but eventually comforting teaching of the creative spirit when he wrote, in the second verse, "'Twas grace that taught my heart to fear, and grace my fears relieved."

One's experience at death's door is likely to be much better for those having had an altruistic bent toward life. The altruistic leaving spirit focuses on our legacies, whatever we achieved materially, socially, and politically, by heeding our promoting, consenting, grouping, role-playing, and help-seeking spirits. Both the physical and spiritual nature of that legacy will be judged by how they have been seen and will be seen to benefit or harm the lives of others—those we helped and those we've hurt. That judgment will essentially define what lives on of our expressed spirits over the generations. The judgment could be bleak. The altruistic expressions that meant so much to our favored groups will mean little to the outside masses. Even worse, the spirits of those expressions will be condemned by those hurt or suppressed by those we've favored. Therein lies the birthing problem when supporting leaders addicted to their ranking, manipulating, and dominating spirits.

Jesus did not disavow our altruistic spirits. In the parable of the talents (Matt. 25:14–30, Luke 19:12–27), he pointed out their importance, but he also elevated our holistic spirits over our altruistic spirits in telling us to give unto Caesar the things that are Caesar's and unto God the things that are God's (Matt. 22:21, Mark 12:17, Luke 20:25). His elevating our holistic desire for oneness over our altruistic desire for social influence is emphasized in each of the first three Gospels in the story of the rich young ruler (Matt. 19:16–22, Mark 10:17–22, Luke 18:18–23). When Martha complained that her sister was listening to his teachings while she was attending the guests, Jesus elevated the holistic spirit's internal injunctions over the altruistic spirit's social dictates in saying, "Martha, Martha, you are anxious and troubled about many things; one thing is needful. Mary has chosen the good portion, which shall not be taken away from her" (Luke 10:41).

More concerning, in two versions of the parable of the wedding feast (Matt. 22:2–14, Luke 14:16–24), he emphasized the possible bleakness of the leaving spirit's judgment at death's door. In both versions, those initially invited were more interested in their work and societal commitments—natural altruistic priorities. Their invitations were withdrawn. The poor and the disadvantaged were then invited, reflecting the host's encompassing spirit. They accepted, possibly because of their feeling so broadly welcomed, although in Matthew's version, one invitee was cast out when he lacked the required wedding garments, garments that might not have been in sync with the encompassing and serving spirits of the celebration.

Contrast the judgment of the disinterested leaving spirit with the deep peace and joy that comes with the holistic birthing spirit. When heeding the holistic desire for oneness with life, we benefit not only those with whom we routinely interact but also whomever comes our way. Our encompassing spirits include them, our serving and truthful spirits never knowingly harm them, and our mindful spirits are ever alert for when that might not be the case. Over time, there comes an inner refrain of a lot of little things of universal and enduring relevance being done well.

That walk is the most natural. Holistic leanings are instinctive in children and can be so with us. Getting lost in the eternal present is a rather simple matter when doing something for the pure joy of it compared to the training, practice, and goal directedness needed by our competitive big-self spirits seeking recognizable success in the ticking now. This is emphasized in Luke 10:21, when "Jesus full of joy through the Holy Spirit, said, 'I praise you, Father, Lord of heaven and earth, because you have hidden these things from the wise and learned, and revealed them to little children.'"

Although the holistic spirits of the little self are simple and straightforward once you choose to heed them, pursuing their long-term rewards rather than the more immediate rewards of their big-self counterparts gave rise to many issues and choices Margaret and Walter faced in these letters. Dad often conveyed where the difficulty lay when telling me, "It's not hard to do what is right once you know in your heart what is right." The importance of learning that sooner rather than later comes through in Margaret's hope for the expressed spirit of one of her students (Appendix 2, 6/4).

Knowing in your heart what is right can be surprisingly hard for so many of us. Jesus noted this on his way to his crucifixion when he told his disciples, "If any man would come after me, let him deny himself and take up his cross and follow me. For whoever would save his life will lose it, and whoever loses his life for my sake will find it." That sake is his holistic orientation.

Those times come when we find ourselves being opposed by economically sensitized crowds easily motivated by opportunistic and egoistically oriented leaders who are fearful of the truthful spirit. In his opposition to Hitler's persuasive rise to power, Dietrich Bonhoeffer demonstrated the freedom and courage that can come from consciously taking that hard walk. His characterization of the choice, recorded in John Baillie's *A Diary of Readings*, was written while he was in prison, not long before his eventual execution:

> We are still left with only the narrow way, a
> way often hardly to be found, of living every

day as if it were our last, yet in faith and responsibly living as though a splendid future still lay before us.

Centuries before the crucifixion of Jesus, Socrates chose that hard walk. It is recorded in the opening dialogue of Plato's *Apology*, with Socrates saying, "How you, O Athenians, have been affected by my accusers, I cannot tell; but I know that they almost made me forget who I was—so persuasively did they speak; and yet they have hardly uttered a word of truth." Later, in addressing them shortly before they voted in favor of his execution, he emphasized the subordination of the big-self desires of our competitive spirits to the little self's holistic desire for the enduring oneness with truth when saying,

> If you say to me, Socrates, this time . . . you shall be left off, but upon one condition, that you are not to enquire and speculate in this way anymore, and that if you are caught doing so again you shall die—if this was the condition on which you let me go, I should reply: Men of Athens, I honor and love you; but I shall obey God rather than you, and while I have life and strength I shall never cease from the practice and teaching of philosophy, exhorting any one whom I meet and saying to him after my manner: You, my friend . . . are you not

> ashamed of heaping up the greatest amount of money and honor and reputation, and caring so little about wisdom and truth and the greatest improvement of the soul? . . . For I do nothing but go about persuading you . . . not to take thought of your persons or your properties, but first and chiefly to care about the improvement of the soul. I tell you that virtue is not given by money, but that from virtue comes money and every good of man, public as well as private. . . . I would have you know, that if you kill such a one as I am, you will injure yourselves more than you will injure me.

Here is a literal translation of that last allegorical statement: I would have you know that if you kill such a spirit as is embodied in me, you will injure your expressed spirits more than you will injure mine. How prophetically Socrates spoke shortly before he took the hemlock rather than deny his soul.

The birthing spirit is key to seeing the vistas of peace and joy that lie ahead for the little self. If, when envisioning death's door, you start hearing something deep inside saying, "Well done, my good and trusting child. You have been faithful over little; now you will be set over much," look around for the vistas that lie ahead.

To get in touch with that voice, try to understand the nature of the kingdom for which Margaret and Walter prayed in their July 5 and July 24 letters. Then, thoughtfully and consistently,

- Heed your mindful spirit when concerned about all that is going on in life.
- Heed your truthful spirit when concerned with what life is disclosing.
- Heed your inclusive spirit when concerned about the worthiness of life.
- Heed your serving spirit when concerned about the many relationships in life.
- Heed your self-searching spirit when concerned with the broken things of life.

Assuring Life in the Hereafter

In Surah 13:20–22, Muhammad is recorded to have said, "They who are true to their bond with God . . . and stay in awe of their Sustainer . . . and repel evil with good. It is these that shall find their fulfillment in the hereafter." This could be paraphrased as follows: those who serve others and are mindful of the creative spirit—and subordinate the expressions of their dominating spirits to those of their inclusive spirits—will find their expressed spirits flourishing in the hereafter. But will there even be a hereafter, and how long might it be?

It won't be long in the astronomical time frame of the ticking now of Brian Greene's *Until the End of Time: Mind, Matter, and Our Search for Meaning in an Evolving Universe*. He writes,

"Planets and stars and solar systems and galaxies and even black holes are transitory." He goes on to say, "For almost five billion years, the sun has supported its tremendous mass against the crushing force of gravity through the energy produced by the fusion of hydrogen nuclei in its core. . . . This standoff . . . will hold firm for about another five billion years . . . [Then] the inward pull of gravity will gain the upper hand. As its spectacular heft collapses inward, the sun's temperature will skyrocket."

Should we be concerned? No! The time frames of the evolution of our starry universe, though fascinating in the here and now of science, make no experiential sense in the generational time frames of the eternal present in which we seek and find meaning. The universal concerns in Table 1 are another matter. They have been around at least since the time of Moses. That's roughly 150 generations ago if we treat a generation as twenty years. That's an "eternity" in human time but not even a blink in astronomical time. No . . . those interested in a thriving world for their expressed spirits do not need to be concerned with the evolution of the starry galaxies in our heavens. They do need to be concerned with the evolution of the plant and animal life on Earth.

Those with a totally egoistic view of life will only be concerned with changes in the immediate present that threaten the desire of their embodied spirits for personal superiority. For them, death is the end, and the judgments of their dying spirits offer only unknown pain and little hope in anything else. Should there be threats, disease, famine, war, fire, or flooding, a person so oriented has clear existential reasons to protect and preserve their personal superiority, but it is folly for them to put their lives at risk for the sake of others.

Those with predominantly altruistic orientations will be concerned with how the evolution of life on Earth will affect the next few generations of those in the groups with which they identify: families, friends, interest groups, businesses and communities, political parties, and those with kindred dispositions. Because our expressed spirits initially and largely lie in our children and their children, these generational concerns inherently appeal to the legacy focus of our leaving spirits. Though these generational concerns are ephemeral in the eternal present, they can be deeply felt and historically relevant.

The significant nature of such generational concerns comes through in the courtship letters of my parents. Having been born into the Greatest Generation, they recalled the Great War; women's right to vote, which ended a century of protest; the Great Depression; and the end of Prohibition. My parents struggled with their acceptance of movies, dancing, and drinking while fearing another depression and another world war with an even more powerful Germany. Their fear of a second depression didn't materialize. Their fear of a second world war did.

It is the role of leaders attuned to their promoting and consensual spirits to develop, propagate, and enforce strategies, policies, and conventions that address the generational concerns of their time. This has certainly been going on since Joseph interpreted Pharaoh's dream as representing seven years of plenty being followed by seven years of famine (Exod. 41). A take on the generational concerns of our time, *Global Trends 2040: A More Contested World*, seen from an American perspective, can be downloaded at https://www.dni.gov/index.php/gt2040-home.

All individuals are variously caught up in the generational concerns of their time. When holistically oriented in our desire for oneness with life in the eternal present, we are freed from the constraints of culture, time, and locality. When concerned with life itself, the mindful spirit places no limit on the relevance of what we might be aware. When concerned with disclosure, the truthful spirit is caringly forthright. When concerned with worthiness, the encompassing spirit recognizes the inherent worth in just being a person. When concerned with relationships, the serving spirit furthers others as they themselves would be furthered. When concerned with brokenness, the self-searching spirit seeks its responsibility in any brokenness in which it plays a part. When concerned with death, your birthing spirit seeks the well-being of all those in which your expressed spirit might reside in the eternal present, which pretty much includes everyone.

Therein lies the little self's joy in this life and the life hereafter. It is realized when bringing joy to others gives you joy. It grows with the assurance that others will share that joy for generations to come. It is magnified with the vision of that joy arriving for everyone when the kingdom of God arrives on Earth. Trying to imagine such a time is not easy. It would be like trying to imagine a time when Jews, Christians, and Muslims jointly anticipate their coming together in Jerusalem to celebrate their mutual joy in the God of Abraham rather than watching the struggle for religious dominance in Palestine.

For such a moment to come in the eternal present, the preceding discussion suggests that, by the time we are adults, we must learn that our competitive big-self desires should not override our desire for oneness with all of life. We must learn along the way that our egoistic desire for personal achievement must not override our altruistic desire for team play. That critically important educational task requires holistically oriented and dedicated teachers with the needed time and resources to carry it out.

Unfortunately, educating students to experience the joy of heeding their holistic spirits counters the interests of politically oriented leaders who are narrowly focused on furthering their desire for social influence or personal superiority. This is obviously the case with egoistic leaders who instinctively heed their manipulating spirits when positioned to control what the public knows and can freely disclose. Given the opportunity, their manipulating spirit will take away not only the freedom of what can be taught but also the right to live from those persons these leaders can convince their supporters to deprecate and despise. We saw this in the mass murder of millions of Jews by the cohorts and avid supporters of Hitler.

It is less obvious but still the case when supporting altruistic leaders whose consensual spirit of disclosure reaches only as far as those they deem worthy. We saw this roughly two and a half centuries ago, when the world's first modern democracy denied women and blacks the right to vote. We saw it again roughly a century later when the Jim Crow laws were promulgated after black men were given the right to vote. We are seeing it today as Bill McKibben so aptly documents in his book *The Flag, the Cross, and the Station Wagon: A Graying American Looks Back at His Suburban Boyhood and Wonders What the Hell Happened*. These are the types of truthful disclosures even altruistically oriented politicians will suppress when their basic appeal centers on getting us to believe that they will make our lives "great."

In the ticking now, it is easy to lose hope that the kingdom our spiritual pathfinders proclaimed will ever come when looking at the power and prevalence of self-serving and oppressive leaders, both past and present. You can, however, have faith that it will come in the eternal present, when looking at the nature of the charge humankind has been given by the creative spirit and at the gift we were given with which to carry out that charge.

Here is an Abrahamic understanding of that gift and the charge as conveyed in Genesis 1:26–28 of the Torah, using the *Inclusive Bible: The First Egalitarian Translation*:

> Then God said, "Let us make humankind in our image, to be like us. Let them be stewards of the fish in the sea, the birds of the air, the cattle, the wild animals, and everything that crawls on the ground."

Humankind was created as God's reflection:

> In the divine image God created them;
>
> Female and male, God made them.
>
> God blessed them and said, "Bear fruit, increase your numbers, and fill the earth—and be responsible for it! Watch over the fish of the sea, the birds of the air, and all the living things on earth!"

Our freedom in interpreting our charge comes through when contrasting the last sentence with its counterpart in the *Jewish Study Bible*: "God blessed them and God said to them, 'Be fertile and increase, fill the earth and master it; and rule the fish of the sea, the birds of the air, and all living things that creep on earth.'" The agreement on the endless creativity of the gift with which to fulfill our charge is seen in the similarity of the latter's translation of the first sentence: "And God said, 'Let us make man in our image, after our likeness. They shall rule the fish of the sea, the birds of the sky, the cattle, the whole earth, and all the things that creep the earth.'"

Our gift of being created in the "image" of the creative spirit of the universe has enabled us to do much in carrying out this charge. Over the past few thousand generations, as time is counted today, we have filled and populated the Earth. A few hundred generations ago, we began documenting our progress as "stewards," "watchers," and "rulers" over the fish in the sea, the birds in the air, and all the living things on Earth. This growing memory is enlarging our charge. In the past tens of generations, we began harnessing and converting energy from the sun and the Earth into ever more useable forms. We can now grow, harvest, and distribute food from throughout the biosphere sufficient to feed us all. We can now understand and heal infirmities and diseases that have long plagued us. And we are now implementing ways of sending news of what is happening throughout the web of humanity in real time. With these uniquely human capabilities, we are becoming the nervous system of life on Earth.

As individuals, we are the "neurons" of this emerging global nervous system (GNS). Via our demeanor, words, and actions, we consciously and unconsciously signal each other in endless ways. A person may tell others what they are suffering, the grieving of a community may draw the world's attention to a locally wrenching event, beholders of the northern lights may share the occurrence, and the interpreters of surveys may alert us to how different communities think about themselves and their environment.

To appreciate the significance of carrying out humanity's role, we have only to look at what our nervous system means to the cells of our body. A single neuron may signal that we are being poked by a pin; neurons in our shoulder may tell us that our arm is being wrenched; visual neurons may bring the delight of a sunrise to mind; and a group of evaluative neurons in our brain may fill us with joy when all is right with our bodies, relationships, and communities or fill us with sorrow when things go wrong.

There are striking dissimilarities as well in how cellular neurons play their individual roles and how we play ours. Cellular neurons variously change how they touch their neighbors but don't move about the body. We like to rove and roam. A cellular neuron plays its role throughout our lives. We play our individual roles for a few decades in the life of humankind, which we hope will be around for hundreds, possibly thousands, of centuries. Our cellular neurons signal each other in time frames measured in hundredths and tenths of seconds. We think and interact with each other in time frames measured in seconds, minutes, and hours.

Just as our minds and thoughts somehow arise out of the interactive signaling of the neurons of our nervous system, it is easy to imagine a "mind" somehow arising out of the expressions we share over the GNS. That mind of humanity too would have "thoughts," but of an order of magnitude more extensive in breadth and time than ours. Could its thoughts be something like the rise and fall of cultures and civilizations? Can we know the "thoughts" of the mind of humanity any more than a neuron can know our thoughts?

The author of Isaiah 55:7–9 suggests not when he writes,

> Let the wicked forsake their ways and the unrighteous their thoughts. Let them turn to the Lord, and he will have mercy on them, and to our God, for he will pardon. For my thoughts are not your thoughts, neither are your ways my ways, says the Lord. For as the heavens are higher than the earth, so are my ways higher than your ways and my thoughts than your thoughts.

That said, this spiritual pathfinder of the Abrahamic tradition clearly felt he was aware of something about the thoughts of the "the Lord" that the "wicked" and the "unrighteous" could not see.

This mental conundrum stems from a difference between the ticking-now time frames of the competitive big-self view of life and the eternal-present time frames of the inclusive little-self view. We see this in the coming judgment by the mind of humanity on the expressed spirits that "the wicked" and "the unrighteous" are birthing if they don't change their ways. That judgment is allegorically expressed in Ezekiel's prophecy regarding the "shepherds of Israel," who were living extravagantly while ruling "harshly and brutally." In that prophesy, the "Sovereign Lord," speaking through the voice of Ezekiel, is castigating a group of the leaders and their supporters for heeding their self-aggrandizing and dominating spirits rather than their mindful and serving ones. Those leaders and their supporters see no need to change their ways as long as all seems well right now.

What we have been able to do in carrying out our charge on Earth reflects our being created in the "image" of the creative spirit of the universe. In her book, *Eve: How the Female Body Drove 200 Hundred Million Years of Human Evolution*, Cat Bohannon suggests how such a mind might emerge. She gives a commonsense and singularly convincing argument on the uniquely important role that maternal and social concerns played in that emergence. She notes,

> Once our climate became unstable—maybe somewhere around *Homo habilis* [~2 million years ago, in the Pleistocene]—the general trainability of our babies' brains would have made a huge difference in their ability to thrive. . . . Human women's brains seem to have evolved a process, unique to pregnant women and new mothers, that helps them adapt to the deeply ancient, ever-challenging sociality that comes with human motherhood, and that this process is neurologically violent. . . . What's unique about the evolution of human brains is fundamentally about our childhoods—that is, our extended period of social learning and the many things our brains do during those periods to optimize living in deep webs of interconnected social groups.

She points out that "no other animal has human grammar. They don't have *language*. They can't cook up complex ideas and dump them into each other's brains simply by swapping around the order of a few sounds" (emphasis in original).

Our genetic instincts for social interconnectedness would have been well in place when *Homo sapiens* became a new species in the animal kingdom roughly two hundred thousand years ago. Not long afterward, we began spreading throughout the continents. The historic charge in Genesis was given somewhere in the middle of the Holocene, marking the end of the last ice age and the beginning of human civilization. There have been times of peace and war in which civilizations have come and gone. All the while, we were finding new ways to shape and exploit our world. The resulting and rapidly increasing impact to life on Earth has given rise to a new designation in our evolutionary time frames that scientists have termed the Anthropocene. Many see it as having started with the great acceleration, which followed the end of World War II.

Our competitive desire to excel in groups with which we identify spurred that acceleration. It has given rise to the social media that enable some to share their expressions at their own time and pleasure though they be miles apart—but it's not enabling all. That freedom will come to all only when we recognize the worthiness of every person regardless of their position on the GNS. It is just a matter of time, but only if *Homo sapiens* is not to end up being nothing more than an abortive thought of Mother Nature, the creative spirit of the plant and animal kingdoms.

So here we are, yearning for peace within ourselves, with others, and with the natural world in which we live. Achieving that peace will not be easy. We must discard the outworn glasses of conventional and prejudicial understandings that give rise to the inequities and conflicts that still trouble us today. We must don new glasses through which to see the unifying hope for our hearts and an enlightened faith in where we are headed. Our need to heed our holistic guiding spirits first and foremost is obvious. But to see where we should head, we need new vantage points from which to view and sort out the issues that must be addressed. We can look at Cat Bohannon's book to see the emergence of the critical role we play in the plant and animal kingdoms. Below, I mention five more books through which to examine the spiritual depth of the problem and its immediacy in the time frames of the GNS and the emergent mind of humanity.

In her book *Doughnut Economics: Seven Ways to Think Like a 21st-Century Economist,* Kate Raworth quickly draws the reader to the heart of the basic economic, social, and ecological issues. She sees the Doughnut Economy as "a radical new compass for guiding humanity . . . [that] points towards a future that can provide for every person's needs while safeguarding the living world on which we all depend." She lists twelve widely recognized needs with which all should be supplied: food, health, education, income and work, peace and justice, a political voice, social equity, gender equality, housing, networks, energy, and water. She also points out a damning inequity:

"The world's richest 1 percent now own more wealth than all the other 99 percent put together." She notes that

> financial income is just one narrow slice of what an economy generates when its aim is to promote human prosperity in a flourishing web of life. . . . Instead of focusing on the throughflow of monetary value, as GDP was designed to do, the new metrics will monitor the many sources of wealth—human, social, ecological, cultural and physical—from which it flows.

In discussing her list of nine ways we over-impact the Earth (climate change, ocean acidification, chemical pollution, nitrogen and phosphorous loading, freshwater withdrawals, land conversion, biodiversity loss, air pollution, and ozone layer depletion), she points out that

> [high-income] countries' global ecological footprints already far exceed Earth's capacity: It would take four planets for everyone in the world to live as they do in Sweden, Canada and the United States, and five planets for all to live like an Australian or Kuwaiti.

The next two books, one Native American and one Celtic, both convey a sense of oneness with nature and with each other that is sorely needed. *Seven Arrows* by Hyemeyohsts Storm describes a time when the homelands of an organically oriented group of hunters, who were guided by their sense of the spirit of nature, struggled with and were overtaken by a materially oriented group of prospectors and farmers, who were guided by their sense of a God of thought. It is a revealing look at the difficulties encountered by the Native American elders trying to preserve the religious traditions and understandings that had given them that needed oneness. Hyemeyohsts Storm's opening comment picturesquely conveys the life-giving understanding:

> The story of these people [the Cheyenne, the Crow, and the Sioux] has at its center . . . the story of the Medicine Wheel.
>
> The Medicine Wheel is the Living Flame of the Lodges, and the Great Shield of Truth written in the Sign of the Water. It is the Heart and Mind. It is the Song of the Earth.
>
> The Medicine Wheel Way begins with the Touching of our Brothers and Sisters. Next it speaks to us of the Touching of the world around us, the animals, trees, grasses and all other living things. Finally, it Teaches us to Sing the Song of the World, and in this Way to become Whole People.

This understanding of being in touch with "the world around us" and becoming whole is central to our fully carrying out our role as the neural system of the Earth. In his book *Sacred Earth, Sacred Soul: Celtic Wisdom for Reawakening to What Our Souls Know and Healing the World,* John Philip Newell conveys a similar struggle that leaders within the Celtic-Christian tradition have had over the centuries in preserving their sense of seeing the creative spirit in both nature and humankind. He points out that "outward authorities need to be read and appraised through the lens of our inner knowing and the deepest experiences of our lives in relation to Earth and to one another." He shares an anecdote in which he began a presentation "using a phrase from the prologue to St. John's Gospel, 'The true light, which enlightens everyone, was coming into the world.'" He writes,

> I spoke of the way the Celtic tradition invites us to look for this light in one another and in everything that has being. Attending the talk that evening was a young Mohawk elder who had been invited to be there specifically to make observations at the end of the talk about the resonances between Celtic and Native wisdom. The Mohawk elder stood with tears in his eyes as he spoke. He said, "As I have been listening . . . I have been wondering where we would be as a Western world tonight, if the

mission that had come to us from Europe centuries ago had come expecting to find light in us.

The last two books emphasize spiritual understandings and actions that are urgently needed. In his encyclical letter *Laudato Si': On Care for Our Common Home*, Pope Francis focuses on the values and goals we should be seeking in our politics and spending. The following statements, with page references, can only give a flavor of his many insights. Their sequencing reflects the purposes of this discussion.

> We need to slow down and look at reality in a different way, to appropriate positive and sustainable progress which has been made, but also to recover the values and the great goals swept away by our unrestrained delusions of grandeur (78).

> We urgently need a humanism capable of bringing together the different fields of knowledge, including economics, in the service of a more integral and integrating vision (95).

> Obsession with a consumerist lifestyle, above all when few people are capable of maintaining it, can only lead to violence and mutual destruction (135).

> The disappearance of a culture can be just as serious, or even more serious, than the disappearance of a species of plant or animal (98).
>
> Discussions are needed in which all those directly and indirectly affected (farmers, consumers, civil authorities, scientists, seed producers, people living near fumigated fields, and others) can make known their problems and concerns (90).
>
> In both urban and rural settings, it is helpful to set aside some places which can be preserved and protected from constant changes brought by human intervention (102).

In his enlightening book *Zen and the Art of Saving the Planet*, Thich Nhat Hanh focuses on how we as individuals must come together in our care for the Earth and for each other. Here is a flavor of his many insights similarly sequenced:

> There is a Zen story about a man on a horse galloping very quickly. At a crossroads a friend of his shouted, "Where are you going?" And the man replied, "I don't know. Ask the horse!" And that is the situation of humanity right now: in our times that horse is

technology. It is carrying us off and it's out of control (170).

Collective awakening is made of individual awakening. You have to wake yourself up first, and then those around you have a chance (12).

An essential condition to hear the call of the Earth and respond to her is silence. If you don't have silence in yourself, you cannot hear her call: the call of life (2).

Can we hear the voices of previous generations, and of the next? Can we hear the voices in our own times that are not being heard? Can we hear the voices of other species, and of the Earth (35)?

New laws and policies are not enough. We need to change our way of thinking and seeing things (12).

We have to come together to do this. We have to take the situation into our own hands. And don't wait for the government: you'll have to wait a very long time (255).

To see what is going on from these vantage points, we must break free of the conventional understandings that we hear from our favorite pundits, pulpits, and politicians. Not that these understandings are wrong, but they too often emphasize limited perspectives from the ticking now of the competitive big self. But once our mindful spirits free us to mount these vantage points by asking ourselves exactly what is being said and why, we will more clearly appreciate what we are being gifted to see in our new role on Earth.

But do we really want to break free from the understandings that have given rise to the inequities and conflicts that still trouble us today? Those of us who do will enjoy sharing the vistas that lie ahead with their children and colleagues. It is in that sharing that we will see the beauty into which our expressed spirits are being birthed.

AFTERWORD

I've often puzzled over my interest in my parents' love letters. I guess life wants examples such as the ones we have here. Can you imagine the difficulty in finding a young man and woman—both with heavy workloads—who would leave a detailed written record of their entire courtship? Life solved that problem by bringing together two potential soulmates whose prior commitments kept them miles apart in trying times while they thirsted for what the other had to say.

Ironically, Margaret and Walter had only a subliminal inkling of what their letters might mean to others. The inkling surfaces from time to time. In her first letter, Margaret hopes that they "might be witnesses for Christ" (*EG*, 17). On December 5, Walter ponders, "Why I write to you these things I do not know, but there are not many with whom I can discuss these problems" (*EG*, 85). On April 3, Margaret pleads, "May God who writes the words have mercy on me" (*QGW*, 270). On May 16, Walter writes, "God certainly moves, works and does in a deep and mysterious way." The following section highlights some of the singular coordinations of the creative spirit related to the publication of their courtship letters.

It was years after that startling Night of the Little Self that Mom and Dad passed away. While sorting through the few belongings they left behind, my sisters discovered and photocopied their courtship letters. I slid my two allotted reams under a bed. A couple of years later, Lois scanned the letters and their envelopes into a PDF file. She sent me a copy on a DVD titled *The Letters of Walter and Margaret*. It was filed away in my computer briefcase.

A year or so later, with nothing better to do, I popped the DVD into my laptop and was soon captivated by finding in the letters answers to a question I had, as a teenager, put to Dad: "Why are you a Christian?" Their letters interrupted my attempting a rather optimistically entitled book, *The Four Pillars of Truth*, with the hope of clarifying some confusion that frequently arises in discussions of the differences between the worlds of the little self and those of the competitive big self. Now in my lap lay a detailed documentation of the emergence and sharing of the spiritual understandings of two seekers struggling with those very differences. Moreover, I was also privy to some very personal consequences of those understandings in their later lives and mine. I briefly set aside my envisioned book to type out the letters and have a few copies printed as *The Letters of Walter and Margaret* for interested members of my extended family.

It was a daunting task. I edited out the more mundane, inserted some orienting comments, and added a news timeline at the suggestion of my niece Claire Barliant, then a fact-checker at the *New Yorker*. I titled the result *Courting God: The Letters of Walter and Margaret*. Now I needed to interest an agent or publisher.

Having coedited a John Wiley book, *Concepts and Applications of Molecular Similarity*, I knew the importance of working with a well-recognized publisher. That book was in a scientific field in which I had been formally trained and had published articles. *Courting God* was in a religious field in which I had little formal training and no publishing experience. Recommended word counts for first-time authors vary but seldom exceeded 50,000 words. Mine had 250,000.

I decided to turn *Courting God* into a trilogy, corresponding to three distinct phases of their courtship: getting to know each other, making sure they really wanted to marry, and planning for their life together after the wedding. *Encountering God*, *Questioning God*, and *Abiding in God* seemed appropriate titles. After I added the Glimpse Back suggestions at the end of each chapter per the editing suggestion of Diane Shepherd, the first volume had roughly 35,000 words.

The enigmatic coordinations of the creative spirit didn't stop there. I attended a fall meeting of the Florida Writers Conference in the hope of finding an interested agent or publisher. One of the attendees suggested I attend the Florida Christian Writers Conference that coming spring, where there would be more interest in the Christian focus of the book.

I arrived at the FCWC on a Wednesday. It troubled me that night that I had lost a little blood as a result of a medical problem I had never experienced before. I went to sleep thinking it would go away. It was still there when I headed for the breakfast buffet.

At breakfast, Margie Houmes took a chair across from me at a long table. After sharing our writing interests, we discussed what sessions and gatherings we planned to attend. When she learned of my interest in meeting agents and publishers, she recommended I sign up for an appointment with Cheri Cowell, who had started EABooks. I was explaining my lack of interest in self-publishing when she spotted Cheri across the room carrying a breakfast tray. With a wave of her hand, she called out, "Hey, Cheri! Come on over."

I don't miss chances to share the uniqueness of my parents' courtship letters when someone is interested. Cheri was. She encouraged me to use one of my three allotted interview times to see the quality of the books she was publishing. I explained my wanting to meet with representatives from more traditional outlets. She understood, but she said she would be free after lunch if I wanted to drop by. Out of kindness and curiosity, I thanked her and said I would.

After lunch, I did. I liked the quality of the books she showed me and appreciated her interest and questions. After briefly sharing our views on the pros and cons of self-publishing, I decided to rest after lunch even though I wasn't tired because of my "little" problem.

That didn't help, but I felt well enough to take advantage of my four o'clock opportunity to sign up for publishing interviews. I lined up meetings with Blythe Daniel of the Blythe Daniel Agency, Diana Flegal of Hartline Literary Agency, and Jim Watkins of Wesleyan Publishing House.

I was looking forward to those meetings the next day as I headed back to my room to get ready for dinner. Seeing the seriousness of my medical problem changed my plans. I needed to make an appointment with my doctor first thing in the morning, but it was a four-hour-plus trip home with no one else to drive. I canceled my three appointments. Jim Watkins kindly prayed with me, and so did Robert Sanders, a former pastor whom I had met that morning.

The symptoms hadn't slackened by the time I went to bed that night. I set my alarm for 7:00 a.m. to arrange an early appointment with Maurice Dees, my doctor, and went to sleep, thanking God that Margie had hailed Cheri over and that "out of the kindness of my heart," I had stopped by after lunch to meet with her. I planned to call her when everything cleared up.

To my utter surprise, after doing my regular exercises the next morning, the symptoms were gone! I had no idea what was responsible for the internal bleeding. It was clearly something that needed to be checked out—but after the conference because I was heading right back to arrange the publication of *Encountering God: Reflections on the Courtship Letters of My Religious Parents* with Cheri.

On the way there, the story of Gideon's fleece in Judges 6:36–40 kept coming to mind. In the story, Gideon looks for a sign that God is really with him. He sets out a fleece on a threshing floor and asks God to make the fleece be the only thing wet with dew in the morning. It turns out to be so. He is still not sure and reverses the test. He asks that all be wet with dew in the morning except the fleece. That too turns out to be so. It gives him the assurance he needs. Unlike Gideon, I didn't put God to the test, but that medical problem was my fleece. It closed a publishing door through which I might not have been able to pass so that I could see one through which I assuredly could.

First thing Monday morning, I met with Maurice. He immediately scheduled an endoscopy. A couple of small polyps were removed, but otherwise, all was clear. I have no explanation for the bleeding other than the creative coordination in which we are all caught up, which freed me to start on the second volume of these letters.

The serendipitous events didn't end there. I was wondering what to do with *The Four Pillars of Truth* when Walter's committing a God-yes logical fallacy in his first letter and Margaret's committing a God-no logical fallacy just two weeks later jumped out at me. Here was a lived lead-in to what I had been wanting to say about the nature of truth. That was soon summarized in an excursus, "Freeing Unifying Religious Truths from Their Partisan Entanglements," which would follow the first chapter. Their writing about the end-times after Hitler's armies entered Austria in March suitably led into a second excursus, "Visions of the Oncoming World Order," following the third chapter. Walter's commenting on the fear of some that "Christ might rise again" and Margaret's mentioning the excitement caused by her engagement ring in their Easter letters led into a third excursus, "Seeing Jesus: Same World, New Eyes," following the fifth chapter.

The puzzling coordinations continued. With the publication of the second volume, *Questioning God's Will: Philosophical Reflections on Pivotal Concerns in My Parents' Letters*, I had begun to see the letters as the way I was being given the words with which to share what I had encountered on the Night of the Little Self. I was still seeking words to clarify my experience of eternal life when I attended another FCWC to promote the first two volumes. There, I was drawn to a talk by Cheri Cowell on the "mission" of a Christian writer. She emphasized the importance of a writer's website. I had deactivated one that Bob Ousnamer at EABooks had created for me in my desire to start a blog related to the notion of tuning forks for joy that was introduced in the first volume.

Now I needed a name more suited to what I was currently trying to convey. The name had to incorporate a sense of the vastness of ones individual expressions, the richness of expressions emanating from a marriage as suggested in these letters, and the breadth of the expressions streaming from our institutions. On thinking of Stephen Gould's notion of science and religion as "non-overlapping magisteria," the notion of truth galaxies came to mind. The website www.ourtruthgalaxies.com resulted.

A personal truth galaxy of traceable expressions made sense as something that, like a legacy, would live on after one's death. However, it lacked the dynamic of the spirits with which I was concerned. Once I turned from what was traceable to what was determining, I had the words I needed. Our embodied spirits are born of all the impressions that flood into us as members of the world relational web; by the same token, our expressed spirits are born of the expressions that flood out from us and spread throughout that web long after our bodies are no more.

The nature of what flows out is largely determined by the guiding spirits we are prone to heed. I found those spirits broadly contrasted in our scriptures: unity versus division, self-insistence versus selfishness, truthfulness versus deception, humility versus superiority, self-responsibility versus faultfinding, generosity versus greed, service versus dominance, immorality versus mortality. I scripturally referenced them in the previous volume (*QGW*, 115–120).

It remained for me to lay down those spiritual contrasts like rows in a spreadsheet with the "good" ones on the right and "bad" ones on the left. But what did the spirits in the same row have in common? It turned out to be the general concern they were set up to address.

One problem remained. Both our spiritual pathfinders and our evolutionary psychologists had raised notions of "good" and "bad." There wasn't complete agreement on either, but especially on what was "good." It involved the difference between our spiritual pathfinder's focus on the inclusive little self's moral priorities in the eternal present and the evolutionary psychologist's focus on the competitive big self's moral priorities in the actual present. Those disagreements were largely resolved by introducing the altruistic column when organizing our elemental spirits by our universal concerns and self-orientations. That completed what became Table 1.

A CLOSING REFLECTION AND INVITATION

When writing "Life-changing experiences like that have to be expressed. But how?" in regard to that Night of the Little Self, I implicitly stated the prompting beneath this trilogy and its last chapter. That "how" first started taking shape in a trilogy having *The Four Pillars of Truth*, *The Little Self*, and *The Coming World Order* as working titles. I was well into the first book when the courtship letters of my parents surfaced. A year later, I felt an urge to make them public in *Courting God: Reflections on the Love Letters of Walter and Margaret*. As a consequence of the interventions noted in the afterword, it soon morphed into a second trilogy with *Encountering God*, *Questioning God's Will*, and *Abiding in God* as working titles.

Although I was not really cognizant of the creative spirit's interest in that trilogy, my immediate objective was clear: to gain a better understanding of the basis of the joy my parents found in life and to further their often-expressed goal of witnessing to their faith in God. The first volume, *Encountering God: Reflections on the Courtship Letters of My Religious Parents*, was a satisfying step forward.

The title of the second volume of letters, *Questioning God's Will: Philosophical Reflections on Pivotal Concerns in My Parents' Letters*, reflects the serendipitous discovery in their letters of life-lived motivations to issues I anticipated raising in the first trilogy. These motivations reflected their interest in spiritual and scientific takes on how things came about, in the realities surrounding the death and resurrection of Jesus, and in the coming world order (or the end-times, depending on your view).

The title of the third volume, *Abiding in God: A Scientist's Insights into Our Guiding Spirits Inspired by the Courtship Letters of His Parents*, reflects the reality, as noted in the afterword, that I and my parents are simply spokespersons for the creative spirit that brought this serendipitous trilogy about. Like me, my parents were simply pursuing their joy in life and that which bound their hearts together in this life and life in the hereafter.

My goals evolved along the way. Reading the Bible and other scriptures following that sudden inversion of my sense of self awakened me to the importance of our guiding spirits and their linkage to our self-orientation. The courtship letters of my parents provided a simple means of demonstrating the variety of concerns that give rise to those spirits. It was in composing the last chapter, "Joy in This Life and the Life Hereafter," that the self-orientation table of our elemental guiding spirits emerged.

That table's role in our spiritual world is usefully viewed as analogous to the role Dmitri Mendeleev's periodic table plays in our chemical world. The paper promoting his table was titled "On the Relationship of the Properties of the Elements to Their Atomic Weights." Replacing "Consequences of the Elemental Guiding Spirits We Heed," the title of the third section of the last chapter, with "On the Relationship of the Consequences of the Appeal to Our Self-

Orientations of Our Guiding Spirits to Their Originating Concerns" would have suggested the analogy.

As with its periodic counterpart, the rows and columns of the self-orientation table are informatively ordered by the stages in life in which they generally have a rising prevalence, the rows from top to bottom and the columns from left to right. And just as elements in the periodic table can come together in complex compounds, our elemental spirits can come together to form complex spirits that can give rise to feelings of love or hate and to notions of freedom or servitude on the part of both individuals and groups.

Clearly, as a consequence of being urged to publish the courtship letters of my parents, I will be singing a new song surrounding that self-orientation table. If you like what is being sung here and would care to join in, please let me know in an email to mark@ourtruthgalaxies.com.

ACKNOWLEDGMENTS

In a book of this kind, it is difficult to recognize all who are to be acknowledged and thanked beyond the Creator and sustainer of the universe and our life on Earth. My parents must be thanked for lovingly passing on their desire to know and abide in God and for saving their letters, in which, as a consequence, I finally found the needed words for what changed on that Night of Little Self. Similarly, my thanks extend to those whose understandings I've quoted and referenced in the body of this work and those mentioned in the afterword. There are groups of individuals who have more directly shaped my takes on the concerns herein addressed and the spirits in which they are addressed. My brothers, Paul, James, and Caleb, and my sisters, Mary, Lois, Ruth, and Anna, principally come to mind for their clarifying recollections of early life on the farm. I want to thank those in the spiritually oriented book groups in which I participated for the creative role they played during the writing of this trilogy: the Bibles and Bagels group at Hodges Presbyterian, then organized by George Ross, with its scriptural focus; the Sack Lunch Theology group at Palms Presbyterian, currently organized by Bob Nelson and Richard Ross, with its theological interest; and the Between the Lines group at Trinity Presbyterian, currently coordinated by Norrie Sanchez, with its commitment to racial, financial, and environmental justice.

There are those who helped bring the book into its current form. My sisters photocopied the letters, and Lois later scanned them into a 1.7 GB file. My niece Claire Barliant suggested the timeline of world events with quotes made possible by the staff at the *Washburn Leader* and the Greeley Public Library. Diane Shepard, editor of the first volume, recommended the self-exploratory questions following each chapter. Kirkus Editorial collaborative editor Keyren Gerlach Burgess suggested a preface and motivating comments to smooth the transitions between the letters and my solo expressions and adding a closing reflection, recommended block quotes in regular type to highlight the letters and ease their reading, and painstakingly flagged every misspelled word, omitted apostrophe, and nonstandard punctuation in the letters that may have been mistranscribed (and some were). Kirkus copy editor Lisa Bannick enhanced my syntax, improved my choice of prepositions, flagged my repetitive phrasings, pointed out needed referencing to earlier letters, and brought my punctuation, italics, and capitalization in compliance with *The Chicago Manual of Style*. Kirkus final polish editor Jon Ford helpfully flagged errors in my transcribing the letters and gave a number of informative editing. I'd also like to thank Rebecca Ford for coordinating the effort on this volume at EABooks Publishing, Monica Miller for her help with the book descriptions, and Bob Ousnamer for his patient help with the cover and the formatting needed to publish the manuscript as a book.

    Finally, I want to thank the many who have encouraged me by putting up with my questions and comments, especially Bill Fleming for his caring and constructive criticism, both grammatical and philosophical, throughout my professional life; Paul and Helene Leaf for clarifying aspects of the Swedish heritage of my grandparents; David Pierce, Lyle Harper, Mark Stanley, and Gerald Leaf for their pastoral thoughts and perspectives; siblings Paul, Lois, and Anna for clarifying discussions on eternal life; friends Scott Matthews and David Brown for their encouraging comments; my granddaughter Chloe and daughter Rebecca for their generational insights; and Martha, my loving life companion, who critiqued the many drafts and significantly influenced the focus while lovingly supporting the drawn-out effort from beginning to end. My thanks to all, especially to her.

APPENDIX 1

A Tally of Walter's and Margaret's Personal Concerns Inferred from the Glimpse Back Sections of this Trilogy, Categorized by Person, Volume, and Universal Concern

The Personal Concern	Margaret				Walter		
	I	II	III	Total	I	II	III

Life

Enjoying life	0	2	1	5	0	1	1
Enjoying marriage	0	0	0	1	0	0	1
Enjoying nature	1	0	1	3	0	1	0
Experiencing God	1	1	1	7	2	1	1
Finding a fit	0	1	1	2	0	0	0
God's guidance	1	0	0	4	1	0	2
Helping others	0	0	0	1	0	1	0
Money	0	1	0	1	0	0	0
The kingdom of God	0	1	1	2	0	0	0

Disclosure

Abiding in God	0	1	2	3	0	1	0
Connectedness	0	0	2	3	0	1	0
Enjoying each other	0	0	0	2	1	1	0
Enjoying scripture	1	1	0	2	0	0	0

Experiencing God	0	0	1	4	1	0	2
Finding a fit	0	0	0	2	1	1	0
God's guidance	0	1	0	2	0	1	0
Helping others	0	0	0	4	1	1	2
Inner peace	0	0	1	4	1	1	1
Their expressed spirits	1	0	0	1	0	1	0
Their relationship	0	1	1	8	0	4	2
Understanding nature	0	2	0	3	0	1	0

Worthiness

Attraction	0	0	1	1	0	0	0
Dancing	1	1	0	2	0	0	0
Drinking	0	1	0	2	0	0	1
Enjoying life	1	0	0	1	0	0	0
Playing cards	2	0	0	2	0	0	0
Politics	0	1	0	2	0	1	0

Relationships

Education	1	0	0	2	0	1	0
Enjoying life	0	0	0	2	0	1	1
Enjoying marriage	0	0	0	1	0	0	1
Helping each other	0	0	0	3	1	0	2
Helping others	0	1	0	3	1	0	1
Their expressed spirits	0	2	0	2	0	0	0
Their relationship	1	2	1	7	1	1	1

Brokenness

Disability	0	0	1	1	0	0	0
Discord	0	0	0	1	0	1	0
Education	0	1	0	1	0	0	0
Financial disappointment	0	0	0	1	0	0	1
Helping others	0	0	0	1	1	0	0
Money	0	0	0	4	0	4	0
Something lost	0	1	0	1	0	0	0
Suffering	0	0	0	1	0	1	0
The country's demeanor	0	0	0	3	2	0	1
The oncoming world war	2	0	0	4	1	1	0
Their expressed spirits	0	1	1	2	0	0	0
Their relationship	0	5	0	9	1	3	0

Death

Their expressed spirits	0	1	1	2	0	0	0

APPENDIX 2

Expressions of Margaret's and Walter's Basic Spirits in the Glimpse Back Sections, Categorized by Their Inferred Personal Concerns[1]

Date & correspondent **Personal concern**	The embodied expressions
	Life: Our Concern for Continued Being
8/19 Margaret **Enjoying life**	If we get a very good kitchen set that is durable and nice looking, a good range and a comfortable bed . . . and a piano, I'll tell you we should live like a king and a queen.
8/28 Walter **Enjoying life**	I had never thot of [the Sabbath] as a gift from God . . . to have . . . as a day to rest and worship and recuperate after six days of work.
9/7 Walter **Enjoying marriage**	Our . . . mating, especially the first night. . . . We need not be [worried about it] if we ask God's help.

---

[1] The bracketed phrases throughout this appendix are mine.

6/3 Walter **Experiencing God**	I do not see how anyone could say there is no God at all, after beholding this earth and all its beauty . . . We who . . . have also all our faculties . . . must certainly thank God without end.
6/6 Margaret **Experiencing God**	John said that I could go along . . . Maybe something will happen so that I can't go but I hope not. "Where He Leads me I Will Follow."
6/15 **God's guidance**	I am glad you prayed as you worked. I often do, even tho it is only in the mind . . . these are often our most sincere and most heard prayers, because they are so spontaneous.
8/18 Walter **God's guidance**	If all the way our Savior leads us, we certainly do not have much to ask beside, do we?
8/25 Margaret **Finding a fit**	Philip . . . asked me to sing a solo. I wonder what song I will be singing. God knows but I shall have to do some searching to know which one it is.
7/5 Margaret **The kingdom of God**	May God lead both of us so that the kingdom may come by us also as we pray in one of Luther's petitions of the Lord's prayer.

9/10 Margaret **The kingdom of God**	May God bring us more closely in love to one another. . . . May we work in His kingdom on earth . . . and then when our summons comes be taken up to be with Him forever.

Disclosure: Our Concern for Truth

6/12 Margaret **Abiding in God**	Philip . . . made it very clear how Jesus' body had to be cut to graft us into Him. . . . God works to cleanse those, who are grafted in, that they may bear more fruit.
5/17 Margaret **Connectedness**	Mrs. Jefferis . . . told me last night about . . . hard financial difficulties . . . [but] was glad that they hadn't waited to be married. Walter, I haven't told her anything about finances.
7/13 Margaret **Connectedness**	Ida just arose and is busy peeling potatoes. She said that I should continue writing. It seems whenever I have something extra to do, I always get help.
5/16 Walter **Experiencing God**	It is hard for us to understand what it means to receive a sudden vision or light, that Paul surely did. I believe Leonard (who is 25) is very sincere.

5/31 Margaret **Experiencing God**	Mrs. Samuel Johnson told Ebba that some years ago there was a group who held street meetings in this town who prayed that there would never be movies shown here. Prayer prevails.
7/14 Walter **Experiencing God**	How important that statement [He is risen, Matt. 28:6] is. If it was not in the Bible or was not true, this world would be vastly different.
5/16 Walter **Helping others**	He felt I was the only one he could talk to and he must talk some
7/21 Walter **Helping others**	A little encouragement and proof of love on the part of a friend, may help to bring out some wonderful talents in another.
5/12 Walter **Inner peace**	I have often wondered at Paul's calm unconcern, in all persecution and affliction. . . .Why? I think it is because of Gods promise and his nearness [Acts 23:11 and Matt. 28:20].
8/15 Margaret **Inner peace**	Reuben sang the song that contains the above verse [Thou wilt keep him in perfect peace whose mind is stayed on thee]. The song continues to ring in my mind's ear.

5/8 Margaret **Their relationship**	Walter, I am so happy that we have a home both an earthly as well as heavenly. . . . I am grateful and glad [knowing] that you will always be helpful, loving, and considerate.
5/20 Walter **Their relationship**	I am determined that we shall take any little differences that may arise between us, to God.
6/19 Walter **Their relationship**	This little sentence [We love, because he first loved us] certainly explains something that I knew but could not put into words, that is; your love for me.

Worthiness: Our Concern for Valued Being

5/23 Margaret **Attraction**	One of the teachers asked me . . . the reason men [those wanting to spend evenings with her] have such a terrible opinion of her. God has spared me . . . such tragic experiences.
7/25 Walter **Drinking**	Mr. Brown has been sober for a long time so I wish you could meet him at such a time. Tho he . . . does not believe in anything that he cannot see or feel, yet he is a nice man.

Relationships: Our Concern for Others

6/14 Walter
Enjoying life

I gather that you are [busy]. . . . There is plenty of work to be done in God's kingdom. Don't you think our tasks would be lighter if we did every one as tho [it were] for Him.

7/12 Walter
Enjoying marriage

I look forward to the day when we may discuss, read, and learn together. Two viewpoints are valuable in any study, don't you think so?

6/28 Walter
Helping each other

Tho I long to see you, do not cut your trip short. I think I can stand it for a while . . . If I only get a letter once in a while.

7/8 Walter
Helping each other

The house is your workshop, where you necessarily . . . spend most of your time, therefore; as much as possible and means will allow, we want it as you want it.

6/28 Walter
Helping others

Jesus makes it very plain here [Mark 10:45] that we are to express our love thru service to one another.

5/11 Margaret
Their relationship

I don't understand in what way I have helped you but I hope that I might always inspire you in some way.

5/17 Walter **Their relationship**	The girls tease you about the hard work but I do not believe it is that bad. If you were here I could help [you] worry . . . as well as encourage you in any way I could.

Brokenness: Our Concern for Rectification

7/23 Margaret **Disability**	During the past few days I have been in bed according to the doctor's orders. . . . Now I have additional time for meditation; God knew my need.
7/24 Walter **Financial disappointment**	I always hoped to have some means when I got married . . . It seems that God did not intend for me to get these means. Perhaps, so that we might strengthen our faith and launch out on him.
7/4 Walter **The country's demeanor**	Today . . . we commemorate . . . our privilege of religious freedom. How do we do it? . . . with a beer bottle in one hand and a firecracker in the other. . . . But enough of this, I am thankful.
9/2 Margaret	Last night I felt so disgusted . . . I looked inward and . . . saw only failures and short

Their expressed comings but as soon as I emptied my heart of
spirits sin on Christ and looked up to Him I found joy.

Death: Our Concern for What of Us Lives On

6/4 Margaret   The funeral of one of my former pupils is to be
Their expressed held this afternoon in Eagle Bend. . . . All of a
spirits sudden he became ill. "We know not what awaits us." I hope he was saved. . . . I am to sing . . . "He knows."

APPENDIX 3

Portrayals in the Gospels of Our Guiding Spirits

Life: Mindful versus Self-Aggrandizing

1. The devil took him to a very high mountain, and showed him all the kingdoms of the world and the glory of them; and he said to him, "All these I will give you, if you will fall down and worship me." Then Jesus said to him, "Begone, Satan! for it is written, 'You shall worship the Lord your God and him only shall you serve'" (Matt. 4:8–10).

2. Whenever you pray, go into your room and shut the door and pray to your Father who is in secret; and your Father who sees in secret will reward you (Matt. 6:6).

3. Take my yoke upon you, and learn from me; for I am gentle and humble in heart, and you will find rest for your souls. For my yoke is easy, and my burden is light (Matt. 11:29, NRSV).

4. And they were bringing children to him [Jesus], that he might touch them; and the disciples rebuked them. But when Jesus saw it he was indignant, and said to them, "Let the children come to me, do not hinder them; for to such belongs the kingdom of God. Truly, I say to you, whoever does not receive

the kingdom of God like a child shall not enter it" (Mark 10:13–15).

5. And one of them, a lawyer, asked him a question, to test him. "Teacher, which is the great commandment in the law?" And he said to him, "You shall love the Lord your God with all your heart, and with all your soul, and with all your mind. This is the great and first commandment. And a second is like it, You shall love your neighbor as yourself. On these two commandments depend all the law and the prophets" (Matt. 22:35–40).

6. When Herod saw Jesus, he was very glad, for he had long desired to see him, because he had heard about him, and he was hoping to see some sign done by him. So he questioned him at some length; but he made no answer. The chief priests and the scribes stood by, vehemently accusing him. And Herod with his soldiers treated him with contempt and mocked him; then, arraying him in gorgeous apparel, he sent him back to Pilate (Luke 23:8–11).

Disclosure: Truthful versus Manipulating

1. God is spirit, and those who worship him must worship in spirit and truth (John 4:24).

2. When the Spirit of truth comes, he will guide you into all the truth; for he will not speak on his own authority, but whatever

he hears he will speak, and he will declare to you the things that are to come (John 16:13).

3. He said to them, "Do you bring in a lamp to put it under a bowl or a bed? Instead, don't you put it on its stand? For whatever is hidden is meant to be disclosed, and whatever is concealed is meant to be brought out into the open" (Mark 4:21–23, NIV).

4. And he came to Nazareth, where he had been brought up; and he went to the synagogue, as his custom was, on the Sabbath day. And he stood up to read; and there was given to him the book of the prophet Isaiah. He opened the book and found the place where it was written,

"The Spirit of the Lord is upon me,
because he has anointed me to preach good news to the poor.
He has sent me to proclaim release to the captives
and recovering of sight to the blind,
to set at liberty those who are oppressed,
to proclaim the acceptable year of the Lord."

And he closed the book, and gave it back to the attendant, and sat down; and the eyes of all in the synagogue were fixed on him. And he began to say to them, "Today this scripture has been fulfilled in your hearing" (Luke 4:16–21).

5. Behold, I send you [his disciples] out as sheep in the midst of wolves; so be wise as serpents and innocent as doves. . . . When they deliver you up, do not be anxious how you are to speak or what you are to say; for what you are to say will be given to you in that hour; for it is not you who speak, but the Spirit of your Father speaking through you (Matt. 10:16–20).

6. The devil . . . was a murderer from the beginning, and has nothing to do with the truth, because there is no truth in him. When he lies, he speaks according to his own nature, for he is a liar and the father of lies (John 8:44).

Worthiness: Encompassing versus Ranking

1. If you love those who love you, what credit is that to you? For even sinners love those who love them. . . . But love your enemies, and do good, and lend, expecting nothing in return; and your reward will be great, and you will be sons of the Most High; for he is kind to the ungrateful and the selfish. Be merciful, even as your Father is merciful (Luke 6:32–36).

2. And they came to Capernaum; and when he was in the house he asked them, "What were you discussing on the way?" But they were silent; for on the way they had discussed with one another who was the greatest. And he sat down and called the twelve; and he said to them, "If any one would be first, he must be last of all and servant of all." And he took a child, and put him in the midst of them; and taking him in his arms, he said to them,

"Whoever receives one such child in my name receives me; and whoever receives me, receives not me but him who sent me" (Mark 9:33–37).

3. Jesus called them [the twelve disciples] to him and said to them, "You know that those who are supposed to rule over the Gentiles lord it over them, and their great men exercise authority over them. But it shall not be so among you; but whoever would be great among you must be your servant, and whoever would be first among you must be slave of all. For the Son of man also came not to be served but to serve, and to give his life as a ransom for many" (Mark 10:42–45).

Relationships: Serving versus Dominating

1. Jesus replied, "A man was going down from Jerusalem to Jericho, and he fell among robbers, who stripped him and beat him, and departed, leaving him half dead. Now by chance a priest was going down that road; and when he saw him he passed by on the other side. So likewise a Levite, when he came to the place and saw him, passed by on the other side. But a Samaritan, as he journeyed, came to where he was; and when he saw him, he had compassion, and went to him and bound up his wounds, pouring on oil and wine; then he set him on his own beast and brought him to an inn, and took care of him. And the next day he took out two denarii and gave them to the innkeeper, saying, 'Take care of him; and whatever more you spend, I will repay you when I come back.' Which of these

three, do you think, proved neighbor to the man who fell among the robbers?" He said, "The one who showed mercy on him." And Jesus said to him, "Go and do likewise" (Luke 10:30–37).

2. When he had washed their feet, and taken his garments, and resumed his place, he said to them, "Do you know what I have done to you? You call me Teacher and Lord; and you are right, for so I am. If I then, your Lord and Teacher, have washed your feet, you also ought to wash one another's feet. For I have given you an example, that you also should do as I have done to you" (John 13:12–15).

3. He looked up and saw the rich putting their gifts into the treasury; and he saw a poor widow put in two copper coins. And he said, "Truly I tell you, this poor widow has put in more than all of them; for they all contributed out of their abundance, but she out of her poverty put in all the living that she had" (Luke 21:1–4).

4. And he told them a parable, saying, "The land of a rich man brought forth plentifully; and he thought to himself, 'What shall I do, for I have nowhere to store my crops?' And he said, 'I will do this: I will pull down my barns, and build larger ones; and there I will store all my grain and my goods. And I will say to my soul, Soul, you have ample goods laid up for many years; take your ease, eat, drink, be merry.' But God said to him, 'Fool! This night your soul is required of you; and the things

you have prepared, whose will they be?' So is he who lays up treasure for himself, and is not rich toward God" (Luke 12:16–21).

Brokenness: Self-Searching versus Faultfinding

1. Then Peter came up and said to him, "Lord, how often shall my brother sin against me, and I forgive him? As many as seven times?" Jesus said to him, "I do not say to you seven times, but seventy times seven" (Matt. 18:21–22).

2. And he told them a parable, to the effect that they ought always to pray and not lose heart. He said, "In a certain city there was a judge who neither feared God nor regarded man; and there was a widow in that city who kept coming to him and saying, 'Vindicate me against my adversary.' For a while he refused; but afterward he said to himself, 'Though I neither fear God nor regard man, yet because this widow bothers me, I will vindicate her, or she will wear me out by her continual coming.'" And the Lord said, "Hear what the unrighteous judge says. And will not God vindicate his elect, who cry to him day and night? Will he delay long over them? I tell you, he will vindicate them speedily. Nevertheless, when the Son of man comes, will he find faith on earth?" (Luke 18:2–8).

3. Judge not, that you be not judged. For with the judgment you pronounce you will be judged, and the measure you give will be the measure you get. Why do you see the speck that is in your

brother's eye, but do not notice the log that is in your own eye? Or how can you say to your brother, "Let me take the speck out of your eye," when there is the log in your own eye? You hypocrite, first take the log out of your own eye, and then you will see clearly to take the speck out of your brother's eye (Matt. 7:1–5).

4. He also told this parable to some who trusted in themselves that they were righteous and despised others: "Two men went up into the temple to pray, one a Pharisee and the other a tax collector. The Pharisee stood and prayed thus with himself, 'God, I thank thee that I am not like other men, extortioners, unjust, adulterers, or even like this tax collector. I fast twice a week, I give tithes of all that I get.' But the tax collector, standing far off, would not even lift up his eyes to heaven, but beat his breast, saying, 'God, be merciful to me a sinner!' I tell you, this man went down to his house justified rather than the other; for every one who exalts himself will be humbled, but he who humbles himself will be exalted" (Luke 18:9–14).

Death: Birthing versus Dying

1. For it [the coming of the kingdom] will be as when a man going on a journey called his servants and entrusted to them his property; to one he gave five talents, to another two, to another one, to each according to his ability. Then he went away. He who had received the five talents went at once and traded with

them; and he made five talents more. So also, he who had the two talents made two talents more. But he who had received the one talent went and dug in the ground and hid his master's money. Now after a long time the master of those servants came and settled accounts with them. And he who had received the five talents came forward, bringing five talents more, saying, "Master, you delivered to me five talents; here I have made five talents more." His master said to him, "Well done, good and faithful servant; you have been faithful over a little, I will set you over much; enter into the joy of your master." And he also who had the two talents came forward, saying, "Master, you delivered to me two talents; here I have made two talents more." His master said to him, "Well done, good and faithful servant; you have been faithful over a little, I will set you over much; enter into the joy of your master." He also who had received the one talent came forward, saying, "Master, I knew you to be a hard man, reaping where you did not sow, and gathering where you did not winnow; so I was afraid, and I went and hid your talent in the ground. Here you have what is yours." But his master answered him, "You wicked and slothful servant! You knew that I reap where I have not sowed, and gather where I have not winnowed? Then you ought to have invested my money with the bankers, and at my coming I should have received what was my own with interest. So take the talent from him, and give it to him who has the ten talents. For to every one who has will more be given, and he will have abundance; but from him who has not, even what he has will be taken away. And cast the worthless servant into the outer darkness; there men will weep and gnash their teeth" (Matt. 25:14–30).

2. When one of those who sat at table with him heard this, he said to him, "Blessed is he who shall eat bread in the kingdom of God!" But he said to him, "A man once gave a great banquet, and invited many; and at the time for the banquet he sent his servant to say to those who had been invited, 'Come; for all is now ready.' But they all alike began to make excuses. The first said to him, 'I have bought a field, and I must go out and see it; I pray you, have me excused.' And another said, 'I have bought five yoke of oxen, and I go to examine them; I pray you, have me excused.' And another said, 'I have married a wife, and therefore I cannot come.' So the servant came and reported this to his master. Then the householder in anger said to his servant, 'Go out quickly to the streets and lanes of the city, and bring in the poor and maimed and blind and lame.' And the servant said, 'Sir, what you commanded has been done, and still there is room.' And the master said to the servant, 'Go out to the highways and hedges, and compel people to come in, that my house may be filled. For I tell you, none of those men who were invited shall taste my banquet'" (Luke 14:15–24).

3. Seeing the crowds, he went up on the mountain, and when he sat down his disciples came to him. And he opened his mouth and taught them, saying:

"Blessed are the poor in spirit, for theirs is the kingdom of heaven.

"Blessed are those who mourn, for they shall be comforted.

"Blessed are the meek, for they shall inherit the earth.

"Blessed are those who hunger and thirst for righteousness, for they shall be satisfied.

"Blessed are the merciful, for they shall obtain mercy.

"Blessed are the pure in heart, for they shall see God.

"Blessed are the peacemakers, for they shall be called sons of God.

"Blessed are those who are persecuted for righteousness' sake, for theirs is the kingdom of heaven.

"Blessed are you when men revile you and persecute you and utter all kinds of evil against you falsely on my account. Rejoice and be glad, for your reward is great in heaven, for so men persecuted the prophets who were before you" (Matt. 5:1–11).

Meet Mark

Mark feels a better understanding of our guiding spirits and concerns expressed in the self-orientation table on the cover of this book can bring a deeper peace and joy in life. To that end, topics for their discussion and related excerpts from the book are being made available at his website www.ourtruthgalaxies.com/your-truth and being promoted on his Facebook page:

http://www.facebook.com/FindYTK/.

He and his wife live in Jacksonville, Florida but often spend time at their vacation home in Hendersonville, North Carolina. He would welcome an email at

mark@ourtruthgalaxies.com

should you or your study group like to meet him and hear more about how our spiritual priorities shape what we experience at the different stages of life.

Made in United States
Orlando, FL
14 February 2025